AN INTRODUCTION TO VEGETATION ANALYSIS

TITLES OF RELATED INTEREST

AN INTRODUCTION TO VEGETATION ANALYSIS

Principles, practice and interpretation

D. R. CAUSTON

*Department of Botany and Microbiology,
University College of Wales,
Aberystwyth*

London
UNWIN HYMAN
Boston   Sydney   Wellington

Published by the Academic Division of
Unwin Hyman Ltd
15/17 Broadwick Street, London W1V 1FP, UK

Allen & Unwin Inc.,
8 Winchester Place, Winchester, Mass. 01890, USA

Allen & Unwin (Australia) Ltd,
8 Napier Street, North Sydney, NSW 2060, Australia

Allen & Unwin (New Zealand) Ltd in association with the
Port Nicholson Press Ltd,
60 Cambridge Terrace, Wellington, New Zealand

First published in 1988

British Library Cataloguing in Publication Data

Causton, David R.
An introduction to vegetation analysis:
principles, practice and intepretation.
1. Botany—Ecology—Mathematics
I. Title
581.5'247   QK901
ISBN 0-04-581024-9
ISBN 0-04-581025-7 Pbk

Library of Congress Cataloging-in-Publication Data

Causton, David R.
An introduction to vegetation analysis.
Bibliography: p.
Includes index.
1. Botany—Ecology—Methodology.   2. Plant
communities—Research—Methodology.   3. Vegetation
surveys.   4. Vegetation classification.
I. Title.
QK901.C33   1987   581.5   87-19327
ISBN 0-04-581024-9 (alk. paper)
ISBN 0-04-581025-7 (pbk. : alk. paper)

Typeset in 10 on 12 point Times by
Mathematical Composition Setters Ltd, Salisbury
and printed in Great Britain by
Biddles of Guildford

Preface

This book has been written to help students and their teachers, at various levels, to understand the principles, some of the methods, and ways of interpreting vegetational and environmental data acquired in the field. Some recent books deal comprehensively with the results of vegetation analyses, for example Gauch (1982) and Greig-Smith (1983), but they are relatively superficial in terms of detailed methodology. Conversely, Pielou (1984) examines the mathematical and computational bases of vegetation analysis in considerable depth, but neglects the methods of acquiring the data in the field and the interpretation of the analyzed results. There are also more advanced books that are unsuitable for a beginner, for instance Whittaker (1973) and Orloci (1978). Here I attempt to survey the whole subject in a largely elementary manner, from the setting up of a survey in the field to the elucidation of the results of numerical analyses.

The methods of vegetation analysis can broadly be divided into two categories: phytosociology, the venerable technique in which methods may be subjective to a greater or lesser extent, and the more objective methods involving a greater degree of mathematical, statistical, and computational sophistication. It is with these latter methods that this book is concerned; phytosociology is mentioned occasionally early on, to enable the reader to make some judgement as to whether phytosociological methods may be preferable for a particular vegetation survey that he or she has to undertake.

The two main types of analysis of vegetation data are classification and ordination; the former groups sample stands of vegetation into relatively homogeneous units, while the latter seeks information about the relationships between individual sample stands. For about two decades after 1959 a plethora of classification and ordination methods were devised, stemming from the pioneering work of Goodall (1953, 1954); but only a few of these became widely popular. Since 1980, the publication of new methods has all but ceased. This is probably because some of the latest methods advanced, particularly in ordination, are efficient and effective. There is now some stability in our subject, and the time seems ripe for a book which could appeal to a broad spectrum of readers, from sixth formers at schools to research workers in plant ecology.

Although very disparate in length, the chapters in this book fall rather naturally into three parts. The first part comprises Chapters 1–3. These chapters contain an introduction to the whole subject, including theoretical considerations and the practical implementation of procedures in the field. There is also a theoretical background to the analytical methods described in later chapters. Chapters 5–8 constitute a second part, dealing with analytical methodology. Chapter 5 is relatively short, and introduces many of the numerical quantities required in vegetation analysis: for example, the χ^2 measure of species association, and Jaccard's coefficient of stand similarity. Chapter 8 is also short, and describes some methods of investigating species–environment correlations. The lengthy Chapters 6 and 7 can fairly be described as the pivotal section of the book, and comprehensively deal with the major subjects of classification and ordination of vegetation data. There are many worked examples based upon a small set of artificial data so that the steps of each analysis can be clearly seen – even for a complicated procedure like Indicator Species Analysis (TWINSPAN). Chapters 4 and 9 comprise the third part of the book, and are devoted to two case studies. These are introduced relatively early in the book (in Ch. 4), partly to break up the more tedious theoretical presentation for anyone reading the book through, and partly so that applications of analytical techniques to real field data can be followed up in Chapter 9 at appropriate places as the reader works through the theoretical material of Chapters 6, 7 and 8.

The two case studies have different origins. The Iping Common transect was a student exercise on a field course, while the Coed Nant Lolwyn study formed part of a larger research project. The background information presented in Chapter 4 reflects this difference. Only the briefest introduction to the Iping Common transect is provided, just sufficient to give some idea of the habitat at the time the survey was carried out. For Coed Nant Lolwyn, however, an attempt was made to reconstruct the history of the site from contemporary records, and the results of the environmental factors measured were compared with those of other published results for British woodlands. These procedures are described in detail, and give a much firmer foundation on which to base the conclusions drawn from results of analytical procedures.

As mentioned above, some of the material in this book should be suitable for use in school sixth forms. Different teachers would no doubt select somewhat different topics as most appropriate for their A-level Biology (and possibly Geography) courses, but much of Chapters 1, 2 and 5 would provide relevant background material for simple field exercises. Programs to execute the simplest classification and ordination methods, written in BASIC, could be run on a typical school microcomputer with the small data sets which would be typically collected in the time available for field work.

In order not to make the book unduly long, I have assumed that the reader has a knowledge of basic statistical ideas and methods, and also of matrix algebra. The latter is only needed for a deeper appreciation of ordination methods, and suitable sources of information are Causton (1983, 1987). A knowledge of statistics is needed for many parts of the book, but mostly at an elementary level. There are a number of suitable texts to satisfy this need.

I am most grateful to my friends and colleagues who have, over many years, both knowingly and unknowingly enhanced my awareness and understanding of topics relevant to this book. To Dr John P. Savidge I owe a particular debt for introducing me to many facets of vegetaton analysis some years ago, and for being a source of inspiration since. Also my thanks to Drs Andrew D. Q. Agnew, Peter Wathern and John P. Palmer for many stimulating discussions about points of interest concerning vegetation analysis in particular, and plant ecology in general. Dr John P. Barkham read most of the manuscript and made valuable comments. Sources of material reproduced herein are acknowledged in the appropriate places, but I must particularly thank here Dr Elizabeth A. Wolfenden for allowing me to use the results of her work in Coed Nant Lolwyn. Finally, my gratitude must be recorded to Mr Miles Jackson, formerly of George Allen & Unwin, for encouraging me to write this book in the first place, and for his continued support through the earlier stages of writing.

Contents

CONTENTS

List of tables

Acknowledgements

We are grateful to the following individuals and organizations who have kindly given permission for the reproduction of copyright material (figure numbers in parentheses):

Figures 2.3, 2.4, 2.5, 5.1 reproduced from *Quantitative and dynamic plant ecology* (3rd edn) by K. A. Kershaw & J. H. Looney with kind permission of Edward Arnold; C. D. R. Centrale des Revues (4.8); Figures 6.6, 6.7 reproduced from *Journal of Ecology* 63, 1975, by permission of Blackwell Scientific Publications Ltd; Figure 7.13 reproduced by permission from R. T. Brown & T. T. Curtis, *Ecological Monographs* 22, copyright © 1966 The Ecological Society of America; Dr W. Junk publishers (7.25, 7.27, 7.28); Figures 7.26, 7.35 reproduced from H. G. Gauch, *Multivariate analysis in community ecology*, by permission of Cambridge University Press; Figures 7.31, 7.32, 7.33, 7.34 reproduced from *Journal of Ecology* 65, 1977, by permission of Blackwell Scientific Publications Ltd; Table 8.4 reproduced from D. B. Duncan, Multiple range and Multiple F Tests, *Biometrics* 11, 1955, with permission from The Biometric Society.

AN INTRODUCTION
TO
VEGETATION ANALYSIS

1

Introduction

Nature of vegetation and analytical approaches

Vegetation is the plant cover of the Earth, and comprises all plant species growing in a very great diversity of assemblages. Before attempting to analyze and understand the structure of vegetation, in terms of the distribution and abundance of the individual species, it is necessary to discuss certain features of the nature of vegetation.

If we look over an area of countryside which has not been too obviously moulded by man, say from a hill-top vantage point somewhere in upland Britain, the panorama of vegetation spread out in front of us appears to be divisible into a number of entities. Thus, much of our imaginary scene may consist of heather moor dominated by *Calluna vulgaris* (heather, ling). Over to the left there is a large plantation of Sitka spruce (*Picea sitchensis*). In the sinuous valley immediately below us, on either side of a small river, is a narrow belt of deciduous woodland in which *Quercus petraea* (sessile oak) is dominant; but a few other tree species, notably *Sorbus aucuparia* (rowan) and *Betula pubescens* (hairy birch), are also frequent. Finally, in a rather wider part of the valley where the soil is obviously much wetter, there is an area dominated by rushes (*Juncus* spp.).

A similar picture of vegetation would be obtained if a larger area of the Earth's surface could be seen, say from an aeroplane. At this sort of scale, vegetation appears to consist of a series of distinct types. On the largest possible scale, namely the whole of the Earth's surface, vegetation again appears as distinct types, for example tropical rain forest, savanna and arctic tundra. These are examples of the great plant formations of the world – plant communities on the largest scale.

If we now descend from our vantage point, enter the woodland in the valley bottom and examine the ground vegetation, things are not as clear cut. We may find large areas of the ground dominated by *Deschampsia flexuosa* (wavy hair-grass), particularly on the steep slopes of the banks; on flatter areas there may be considerable stands of another grass, *Holcus mollis* (soft fog). Here and there are small stands of bracken (*Pteridium aquilinum*) and, in other places, small thickets of bramble (*Rubus fruticosus*). So our first sight of vegetation on a smaller scale, in a much

1

more restricted area, still gives the impression of distinct vegetation types or plant communities. This notion might be thought to be reinforced by examination of each supposed community, when associated species may appear to be restricted to communities defined by the above species. For example, *Galium saxatile* (heath bedstraw), *Teucrium scorodonia* (wood sage) and the moss *Polytrichum formosum* might accompany *Deschampsia flexuosa* and not be found elsewhere in the area; *Oxalis acetosella* (wood sorrel) may be quite common in the area dominated by *Holcus mollis*. However, bluebells (*Hyacinthoides non-scripta*) might occur to a greater or lesser extent in the *Holcus*-, *Pteridium*- and *Rubus*-dominated areas, while elsewhere forming almost pure stands themselves.

When one comes to make a really close examination of an area of ground, in particular by laying small quadrats and closely scrutinizing the make-up of the vegetation therein, it is then realized that a simple classification of vegetation into plant communities can scarcely be realistic. For instance, if one lays a sequence of contiguous (touching) quadrats along a transect (straight) line across a bracken-dominated zone, starting and ending well away from any bracken, it is extremely difficult to delimit the bracken area itself. In the middle of the clump, the bracken may be so dense as to entirely suppress all other species by its shade in the summer and the density of its litter at other times of the year. On either side, the density of the bracken becomes less and less; but it will be found quite impossible to say exactly where the bracken zone ends, because over quite a length of the transect, away from the area of dense bracken, many quadrats will contain isolated bracken stems and some quadrats may contain more than one. Such quadrats may even be separated from the main bracken clump by one or more quadrats devoid of this species.

For many years, plant ecologists were divided in their opinions as to whether vegetation consists of a series of distinct communities or whether vegetation types grade into one another, i.e. vegetation is a continuum. It would seem that the argument can be resolved quite simply by recognizing that on a small scale vegetation is a continuum; but on a larger scale, where the vegetation is not scrutinized too closely, a community structure can be recognized.

Even those ecologists who favoured the community idea of vegetation recognized that boundaries of communities are indistinct. They coined the word **ecotone** to describe the area of the boundary, and the ecotone is considered to be a distinct vegetation type in itself. Very often the ecotone is found to be more species-rich than either of the communities it separates, and this is ascribed to the fact that not only does the ecotone contain species of both adjacent communities, but that it has species which occur in neither community. In other words, the ecotone could really be recognized as a community itself.

2

These facts are very easily accommodated within the continuum idea of vegetation. If we imagine a transect from the centre of a clump of bracken out into an area largely dominated by *Holcus mollis*, then we can say that for equal distances traversed along the transect the continuum changes slowly in the bracken area, changes rapidly at the boundary and then changes slowly again beyond the boundary. Most ecologists now recognize that small scale vegetation structure is a continuum.

The above discussion is of central importance to the topics in the remainder of this chapter.

Purposes of vegetation analysis

There are basically two reasons for surveying and analyzing vegetation. One of these is for description and mapping purposes, and the other reason is for what we shall call 'ecological' purposes. Since the prime objective of ecological enquiry is to determine the factors, both biotic and abiotic, which control the occurrence and distribution of plant species, the phrase 'for ecological purposes' will mean the use of vegetation analysis to investigate species–species and species–environment relationships.

It seems convenient to classify the various aims of vegetation analysis into three categories, based on the scale of the work in the field:

(a) large-scale vegetation survey, usually of a new area, for description and mapping;
(b) small-scale survey of a restricted area, containing different vegetation types, where the objective could either be mapping or ecological purposes;
(c) more detailed work after (b), which might involve comparisons between the different vegetation types, or more detailed work on individual species found in the whole area.

Essentially, different methods are optimal for these three categories of usage, but to some extent methods best suited to one purpose may be used for others – it is up to the user to choose.

Large-scale survey

This kind of survey is made when interest centres on describing and mapping the vegetation of a large area. At this level of working, distinct communities may be recognized, and the aim is almost invariably a description and classification of community types. For this kind of work the methods of phytosociology (p. 4) are particularly useful, and it is probably true to say that most large-scale surveys employ some kind of phyto-sociological approach.

Small-scale survey

At this level of working it is usually very difficult to adhere to the community concept, for the reasons discussed in the first section of this chapter. Here the continuum nature of the vegetation must be recognized. As we shall see later (p. 35), although classification is not the most appropriate way of dealing with a situation of continuous change, there may be reasons for wishing to classify the vegetation. This will certainly be true if the prime purpose of the survey is the production of a map, and a classification also facilitates vegetation description. If the purpose is ecological, then the more appropriate technique of ordination (p. 35) is much better; but even for ecological objectives, classification may be useful if only because the overall vegetation of the area is broken down into more manageable 'chunks'. However, phytosociological methods of classification are scarcely useful in this context — more objective methods are required.

More detailed field work

Here we are really out of the realm of general survey, and the overall analytical techniques of classification and ordination are usually not applicable. Indeed, it is impossible to generalize; the methods of field work and analysis selected will depend on the job in hand. However, it can be said here that under this heading we will be doing field work of the most precise, and perhaps most tedious, kind. Furthermore, the methods of sampling and the subsequent statistical analyses will also have to be especially rigorous. Some guidance to methods available, and the problems involved, is given in the second half of Chapter 2.

Phytosociological and more objective methods

Phytosociology is defined as the science of vegetation; it does not primarily concern itself with the interaction between vegetation and environment, except in the more qualitative sense. The aim of phytosociology is a world-wide classification of plant communities. Various schemes have been put forward but the best known one, at least in western Europe, is that due to Braun-Blanquet (1927, 1932, 1951) and known either as the Braun-Blanquet School, or the Zurich–Montpellier School, of Plant Sociology. An account of the methods involved have been summarized by Kershaw &

Looney (1985), and earlier Poore (1955a, 1955b, 1955c, 1956) gave an extensive description of the approach together with a critical discussion of the principles.

The science of vegetation, phytosociology, deals with floristic structure, development, distribution and definition of plant communities. This is distinct from plant ecology, which also investigates the reasons for all the above features of plant communities, and so is concerned with all the physical and biotic factors of the habitat and not just with the plant community itself. This divorce is not liked by British ecologists, and phytosociology has been unpopular in Britain. On the European continent, the origin and popularity of phytosociology arises from a preponderant interest in classical taxonomy and phytogeography. Plant communities are not only classified into primary groupings – called **associations** – but associations are grouped into higher categories analogous to a taxonomic ranking of species into genera, families, orders and classes. Thus in the Zurich–Montpellier School of Phytosociology associations are grouped into **alliances, orders** and **classes**; and, in theory at least, any plant community can be classified in the same way as can an individual plant. On the other hand, in Britain and the North American continent interest has centred on plant ecology, as defined above, and so ecologists in these countries have been largely out of sympathy with the aims and methods of phytosociology.

As stated in the previous section, phytosociological methods are useful in large-scale vegetation surveys, and if one is working in a large new area they may be the only practical methods to apply. In this book, however, we are primarily interested in vegetation analysis for ecological purposes, and in these circumstances phytosociology finds little place. One reason for this is that such methods are highly subjective, and subjectivity is useless for the statistical methods that need to be applied when investigating species–species and species–environment relationships.

Samples

It is obviously quite impossible to consider all the plants making up vegetation unless interest centres on an extremely small area; in general, some kind of sampling scheme is required. Broadly, there are two kinds of sample – stands and plotless – and we shall describe each in turn.

Sample stands

A sample stand is delimited as a definite area of ground containing a stand of vegetation. There is no limit on size or shape, it depends on the purpose in hand; but we can generalize to some extent according to whether we are using phytosociological methods, when the sample stands are called **relevés**, or more objective methods, when the term **stand** may be used.

RELEVÉS

A relevé is a *carefully chosen* sample stand in what the phytosociologist considers to be a homogeneous stand. Homogeneity has been defined by Dahl & Hadač (1949) in the following way (extract from Kershaw & Looney 1985):

> A plant species is said to be homogeneously distributed in a certain area if the possibility to catch an individual of the plant species within the test area of given size is the same in all parts of the area. A plant community is said to be homogeneous if the individuals of the plant species which we used for the characterisation of the community are homogeneously distributed.

In essence, this definition contains two difficulties. First, that certain species characterize the community, and these have to be selected first. Secondly, the definition implies that these species occur on the ground at random, but there is overwhelming evidence that this state of affairs rarely occurs. Recognizing these difficulties, Dahl & Hadač further state (extract from Kershaw & Looney 1985):

> In nature plant communities are never fully homogeneous … We may thus talk of more or less homogeneous plant communities. Measurements on the homogeneity of the plant community are rarely carried out, but the human eye, badly adapted to measurement but well to comparison, rapidly gives the trained sociologist an impression whether a plant community he has before his eyes is highly homogeneous or not.

This last sentence epitomizes the subjectivity of the process.

The size of a relevé can be arbitarily selected by the phytosociologist, and will typically be as large as possible but lying wholly within the area of vegetation considered to be homogeneous. A large size is necessary so that as many of the rarer species of the community can be included as possible.

STANDS

A stand of vegetation used as a sample unit must be delimited, and this is usually done by laying a **quadrat**. To many people, a quadrat is typically thought of as a square area, up to 1 m^2, defined by a wooden or wire frame. However, there is no reason why a quadrat should be square, although it almost invariably is. Quadrat size is a very tricky question, and depends on the purpose in hand. More will be said at appropriate places in the book (pp. 21 & 81) but, in general, a quadrat size up to 0.25 m² is suitable for herbaceous vegetation, while very much larger sizes than 1 m² are required for work with most woody species. Such quadrats must be laid out with string and pegs, which is, of course, very tedious.

Plotless samples

For certain purposes one can dispense with sample stands and simply define sample points in the study area, from which various measurements can be made. Although plotless sampling methods can be employed in any vegetation type, its most common use is when dealing with trees, and obviates the necessity for laying out very large quadrats. Full details are given on pages 28–31.

Types of data

When using vegetation analysis methods for ecological purposes, two kinds of data are normally required – vegetation and environmental. Either sort of data can be qualitative or quantitative in form, and we shall discuss the various types below.

Vegetation data

QUALITATIVE

Qualitative, or **presence/absence**, data are simply a list of the species occurring in each stand. They are much the easiest kind of data to gather in the field since a species is scored as soon as it is found, without then having to look for it over the whole area of the stand.

QUANTITATIVE

As before, this form of data consists of a species list in each stand, but now an abundance value of some kind is added. There are many scales of abundance in common use: some which must be actually measured; others which can either be estimated by eye, or are based on some other qualitative judgement.

An old, and highly descriptive, scheme is that of frequency symbols. A species is scored with one of the following terms: dominant (d), abundant (a), frequent (f), occasional (o), rare (r). This approach has been widely used in the past when a study area, such as the whole of a small wood, was examined completely and not by means of sample stands (see, for example, the habitat studies in Dony 1967). Under such conditions, another term − local (l) − was often added as an adjective to give, for example, locally abundant (la). Obviously frequency symbols are not applicable to small stands, but might on occasion be useful for large stands or for relevés.

Another, more numerical, estimate by eye is based on the idea of species **cover**; that is the percentage of the stand area occupied by aerial parts of the species. Cover is essentially a vegetation parameter which can be estimated by measurement (p. 19), but it can also be estimated much more easily by eye; the actual measurement of cover in the field is extremely laborious.

Estimating cover by eye to the nearest 1% is almost out of the question, but it is more feasible to the nearest 5%. However, if the cover of a species in a stand is estimated to be less than 5%, then a real attempt should be made to 'get it down' to the nearest 1%. If these recommendations are followed, then the visual estimation of species percentage cover is very useful as a crude and subjective measure of abundance in general survey work, but it should not be used in more detailed studies.

There are, however, a number of 'scales' or ratings based on cover which also have had wide use in the past, and three of these, the two Braun-Blanquet ratings for relevés and the **Domin scale**, are given in Tables 1.1–1.3. For beginners, a simpler scale for estimating cover, with fewer points, is better; but with only few points it is essential for the scores to represent approximately equal intervals on a logarithmic rather than a

Table 1.1 Braun-Blanquet cover scale.

Rating	Description
+	sparsely, or very sparsely present; cover very small
1	plentiful, but of small cover value
2	very numerous, or cover 5–20%
3	any number of individuals; cover 25–50%
4	any number of individuals; cover 50–75%
5	cover greater than 75%

Table 1.2 Braun-Blanquet grouping scale.

Rating	Description
Soc 1	growing singly; isolated individuals
Soc 2	grouped or tufted
Soc 3	in small patches or cushions
Soc 4	in small colonies, in extensive patches or carpets
Soc 5	in pure populations

Table 1.3 The Domin scale.

Rating	Description
+	isolated; cover small
1	scarce; cover small
2	very scattered; cover small
3	scattered; cover small
4	abundant; cover about 5%
5	abundant; cover about 20%
6	cover 25–33%
7	cover 33–50%
8	cover 50–75%
9	cover 75–under 100%
10	cover about 100%

Table 1.4 Short, logarithmic scale.

Rating	Description
1	cover up to 2%
2	cover 3–10%
3	cover 11–25%
4	cover 26–50%
5	cover 51–100%

linear scale, because small differences among small cover values are of greater significance than similar differences among high cover values. Such a scale, of only five points, is given in Table 1.4.

A vegetation parameter that is actually measured seems to be less usually employed in survey work; but there is no reason why such a parameter should not be used thus, and an example is given in one of the case studies (Ch. 4). It always takes longer to measure than to visually estimate and so fewer stands can be sampled in a given time. This is probably the reason for the preference of visual estimates over measurement in primary survey work, where maximizing the number of sample stands is more important than any increase in accuracy of abundance assessment.

Environmental data

QUANTITATIVE

Most environmental data involve a measurement of some kind and so are inevitably quantitative in nature. Typical environmental factors measured

9

in stands are those of the soil, which involve taking a sample back to the laboratory. It is, however, becoming increasingly possible to make soil measurements in the field by means of various probes connected to portable electronic equipment; but either the soil must be very wet, such as in a peat bog, or the soil must be wetted sufficiently to enable the probe to be inserted. Apart from pH probes, other probes are expensive and have short lives; this, and the necessity for taking a lot of deionized water into the field, normally restricts soil analysis to the laboratory.

Many other environmental factors can be measured besides those of the soil. The topographic features of slope and aspect can be given for each stand. Slope can be measured either with a clinometer, or by levelling each stand by dumpy level and staff and plotting the profile(s) if a transect or grid method of sampling is used. Biotic features, such as the number of rabbit faecal pellets, could be relevant, so also could the general height of the vegetation in the stand. Soil depth may be important. It all depends on the precise features of the particular area under study: it is no good trying to count rabbit pellets where there are no rabbits, and soil depth would be irrelevant and impossible to measure in a fen!

As far as the aerial environment is concerned, temperature, light, and humidity are important factors. The problem here is one of instrumentation, which is expensive, and ideally one requires a similar instrument in each stand. Several ingenious inventions have been devised to overcome the need for expensive standard instrumentation. Thus for light intensity, which is particularly important in woodland, an integrated light measurement over a defined period is required. Normally this would require a selenium cell connected to an integrator in each stand – exorbitantly expensive – but Wolfenden *et al.* (1982) have developed an inexpensive photochemical light meter for precisely this purpose and employed one in each of 200 stands in a woodland survey (Case Study no. 2).

QUALITATIVE

Occasionally items of environmental data are naturally qualitative. For instance, Blackman & Rutter (1946) in an investigation of the distribution of *Hyacinthoides non-scripta* in relation to light intensity scored each stand according to whether it was under a larch (*Larix*) canopy (1) or not (0). On chalk grassland it is important to note whether the stand includes the whole or part of an ant hill (1) or not (0). Other environmental factors of this nature may be relevant, and they have the advantage of being particularly easy to record in the field.

2

Field methods

Although, to some degree, any field method can contribute to either of the two main purposes of vegetation analysis identified in Chapter 1 – descriptive and ecological – and most authors in the past have simply lumped all such methods together leaving the user to sort things out, it is much better in a systematic survey of field methods to deal with those that are more suited to descriptive and ecological purposes separately. This is not to say that there must be a rigid demarcation, but I think that this kind of partitioning is helpful to the student to find his or her way through a number of theoretical concepts and practical approaches towards the method best suited for the precise aim of a particular study.

This chapter will, therefore, be divided into two parts. Methods particularly suited to primary survey, in which the aim is to record all species present (or at least, say, all species of flowering plants and ferns), will be presented first. The important feature of these methods is that many species are involved. For more detailed study, attention is usually confined to fewer species and rather different methods ideally apply.

Primary survey – many species

The first decision to be taken when embarking on a vegetation survey is whether to sample in a random or regular manner. This means that, having selected the survey area, are the stands to be delimited at randomly chosen points or is there to be a regular relationship between the stand positions? The merits and disadvantages of each approach will be discussed after describing the methods.

Methods of random sampling

Basically, the idea is to define a pair of co-ordinate axes in relation to the area, as shown in Figure 2.1. A study area could well be delimited rectangularly, but there may be good reasons for it to have an irregular outline if the boundary of the area is following some natural feature, for example a wood edge, or parallel to a cliff edge allowing for a margin of

11

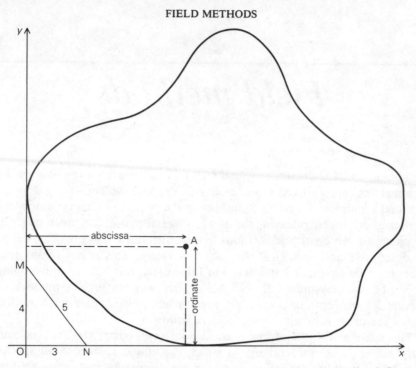

Figure 2.1 A pair of coordinate axes in relation to a study area outlined. See text for explanation of the right-angled triangle near the origin.

safety! If the outline is irregular, the co-ordinate axes must be sufficiently extended to embrace the extremities of the area, as in Figure 2.1.

In a small area consisting of low vegetation, the axes can be physically represented by measuring tapes. In placing the tapes, the x-axis could be placed first using a compass to specify its orientation, and then the y-axis laid down making sure that it is properly at right angles to the x-axis. Again, the compass may be used to ensure a right angle between the two axes; but if the terrain is fairly smooth and the vegetation low, then a third tape can be used to form a temporary right-angled triangle (MNO) of sides 3, 4 and 5 units, as shown in Figure 2.1. These side lengths accord with the result of Pythagoras' theorem for a right-angled triangle, i.e. $(MN)^2 = (MO)^2 + (NO)^2$ so $(5)^2 = (4)^2 + (3)^2$. This is more accurate than using the compass if the terrain allows. Having laid the axes, random points, such as A, are found by using random number tables or, better still, by generating random numbers on a scientific pocket calculator. For each point a pair of random numbers is required: the first representing the x-co-ordinate or abscissa, and the second number designating the y-co-ordinate or ordinate. Then a tape may be run out from the abscissa value on the x-axis, and a similar one from the ordinate value on the y-axis (dashed lines in Fig. 2.1) and the random point will be at the point of intersection of the two tapes when they are each run out to the required distance.

If several people are working in the area, such as students on a field course, each defining and seeking random points, this procedure will result in a large number of criss-crossing tapes which are constantly being moved, and everyone will get in each other's way. An alternative is to lay out a grid, perhaps of strings marked at every metre if the number of tapes becomes prohibitively expensive. An example is shown in Plate 1, where individual students on a field course are locating random points, laying quadrats and recording therein. The squares of the grid can be any convenient size, and once a random pair of co-ordinate values has been generated the location of the point can be immediately identified within a particular square and the final position determined by measuring in the x- and y-directions only within that square. Even ordinary 30 cm rulers might be sufficient for this.

For larger areas of open country, a grid of ranging poles could be established and then tapes used within each square. In this case using compass bearings, perhaps combined with sightings on objects identifiable on large-scale maps, is the most accurate way of defining right angles and setting up the grid lines in general.

In woodland a different approach is required, as it will not usually be possible to lay out tapes or strings in straight lines, or sight ranging poles, over anything but short distances. If an aerial photograph of the study area is available, it may be possible to use it to draw an accurate large-scale plan of the outline of the wood. If such a photograph is unavailable, one may have to rely on an enlarged copy of a large-scale map; but a difficulty here is that such maps are often based on old surveys and the boundaries of woodland could well have changed. However it is done, once you have a large-scale plan of the woodland boundaries the next job is to go out to the wood and obtain distances and compass bearings of a number of large trees from boundary features that can be plotted on the map. The positions of these marker trees can then be transferred to the map. It is advisable to number the marker trees, both in the field and on the map. By using random numbers, stand positions can be obtained, as previously described, and marked on the map. The position of each point, in terms of both distance and bearing from the nearest marker tree, is determined from the map and finally, when these measures are transferred to the wood, the position of the random point on the ground is found.

PROBLEMS OF REPETITION AND OVERLAP

In putting the above procedures into practice, two related final problems need to be resolved: sample repetition and sample overlap. The first is conceptually simple: in a run of random number pairs there is a small probability that a particular pair will be repeated. If we had decided to have

n stands in our sample, we obviously require n different samples, so we must be careful to eliminate any repetitions and continue generating random numbers until we have n distinct pairs. In statistical terms, where sampling theory is often described in relation to selecting objects out of a bag, the above avoidance of repetitions is known as 'sampling without replacement'; in other words, once an object is selected it is not returned to the bag where it might be sampled again.

The second problem, of stand overlap, arises because random numbers are essentially a continuous sequence, whereas placings of, say, a square quadrat of side 0.5 m would have to be at least this distance apart to avoid overlap. The procedure adopted to avoid overlap and repetition is most conveniently described by means of an example.

Suppose we have a rectangular area of ground 20×10 m, and are sampling with a square quadrat 0.25 m^2 in area (i.e. one having 0.5 m sides). There are 40 possible non-overlapping quadrat positions along the longer (x) axis and 20 such positions along the shorter (y) axis; thus, the total number of possible non-overlapping stands in the area is $40 \times 20 = 800$, and the unit of sampling in each direction is 0.5 m. If we are using random number tables, we require numbers going up to 19.5 in the x-direction (i.e. three digits) and 9.5 in the y-direction (i.e. two digits). In the x-direction, the first of the three digits can only be 0 or 1, all other trios of random numbers must be discarded; but in the y-direction all pairs of digits can be used. However, in each case the final digit must be rounded up or down to 0 or 5, that is digits 8, 9, 0, 1, 2 would be recorded as 0 and digits 3, 4, 5, 6, 7 recorded as 5. This procedure would avoid overlaps. If we wished to sample one-tenth of the whole area, 80 stands would be required. The best procedure would be to select 160 appropriate random numbers, three-digit and two-digit alternately, rounding up or down as we go; we should then have 80 non-overlapping quadrat positions, but probably some repetitions. Finally, by selecting more random number pairs, the repetitions could be eliminated.

If using a random number generator on a calculator, typically the result will lie between 0 and 1. For the x-axis we require the first three digits, discarding all those not beginning with 0 or 1, and for the y-axis the first two digits are used. A programmable calculator would be ideal: not only could it be programmed to do the roundings of the final digit and eliminate out-of-range numbers but, with sufficient memory for storing the sequence of generated numbers, it could also eliminate repetitions.

Regular sampling – transects and grids

If it is desired to survey the whole area by taking regularly placed sample stands, then a two-dimensional sampling scheme must be utilized (a grid). If

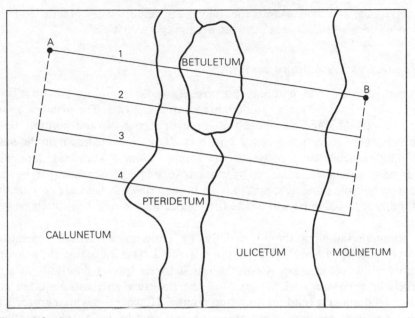

Figure 2.2 Diagrammatic representation of part of an area of heathland, having a series of vegetation types down a slope (low ground on the right), together with a superimposed grid of transect lines.

the main vegetational changes are essentially unidirectional, as in Figure 2.2, then a single transect line (AB, as shown) may be sufficient for an initial description; but if a map of the area is required then a grid (of transects), as shown in Figure 2.2, is needed. Another way of viewing a grid sampling method in the context of a mainly unidirectional vegetation change is that the different transects can, to some extent, be regarded as replicates; not entirely so, since there will almost always be some small but real differences within each vegetation zone. An obvious example exists in Figure 2.2, where transect 1 crosses the widest part of an area dominated by birch trees (Betuletum), transect 2 intersects a narrow part of the Betuletum, whereas transects 3 and 4 do not traverse the birch area at all.

A question that does not arise in a random sampling scheme is, how far apart should the stands be in the transect or grid? For any particular survey the question is usually answered indirectly. First the length of the transect or transects is decided upon according to the number of vegetation types it is desired to include. Then the number of transects and their distance apart are selected on the basis of the total number of stands that it is considered possible to examine in the time available, bearing in mind that it is desirable to have the stands the same distance apart in both directions, that is within and between transects. For very small-scale surveys, stands may be

15

contiguous; and if the stands themselves are small, a very precise description of vegetation changes over the area may be obtained.

Regular versus random sampling

For description and mapping purposes, regular sampling is virtually a necessity, and there are no disadvantages in doing this. The distance apart of the sample stands is under the surveyor's control, and contour lines showing approximate vegetation type boundaries may be drawn on the map joining, and/or going between, the sample points. The field data will normally be classified, either by a phytosociological approach or by the more objective numerical methods to be described in Chapter 6; and the stand groups obtained will define the vegetation types to be plotted on the map.

Random sampling should, strictly, be undertaken when a vegetation survey is done with ecological purposes in mind. This is because the primary aim is to detect and assess correlations between species distributions and levels of environmental factors, and the statistical methods required are only valid under a random sampling regime. However, having perused the two previous sections, you are probably rightly aware that random sampling is more time consuming in the field and more complicated to achieve than using a regular sampling method. Hence the question may be asked, is it *really* necessary to sample vegetation randomly in order to perform a valid statistical analysis of the results?

Whichever way one samples, the selection of the study area and the establishment of an origin and direction for a transect (or a two-way regularly or randomly sampled grid) must always be subjective, unless one establishes these bases by objective random methods on a large-scale map, and then transfers the origin and direction to the corresponding positions on the ground. Unless this is done, the completely random element is removed from a sampling scheme.

Consider now that a transect has been established across some obvious vegetation and/or environmental gradient, and that stands are to be examined at regular intervals along the transect. Obviously, in relation to this particular gradient sampling is not random; but in relation to many of the other multitude of types of change of less pronounced vegetation types or environmental factors such a sampling scheme would be effectively random since no conscious choice could be made by the investigator in relation to these other factors, as their nature and positions in the study area would be unknown. Thus, if the subjective choice in positioning the transect or grid is based on as few obvious features of the vegetation and/or environment as possible, regular sampling *could* be a viable alternative to random sampling for statistical tests to remain valid.

16

Qualitative or quantitative data?

It must first of all be recognized that qualitative data are more easily and rapidly acquired than are quantitative species data, even when the quantity is only estimated by eye rather than by measurement. Hence, if as many stands surveyed as possible is the prime criterion in the work, data must be restricted to species lists in the stands. In large-scale primary survey using phytosociological methods, Braun-Blanquet cover and grouping scales (Tables 1.1 & 1.2) are required for each species in a relevé. For large- or small-scale primary surveys using quadrats for mapping, qualitative data should be suitable.

For most other work it is worthwhile to collect quantitative data. This particularly applies to the commoner species in the data set; but since it is not always apparent before the field work is carried out which *are* the commoner species, it means that all species should be quantitatively assessed. Although, as we shall see, only qualitative data are required for classifications and may also be preferable for ordinations of primary survey data, quantitative records can provide valuable additional information (see the case studies). Estimation of cover values by eye and expression in terms of the scales given in Chapter 1, or as a percentage, is by far the commonest method, but any of the vegetation parameters described in the next section could be used.

More detailed field work – one or a few species

Stratified random sampling

Let us assume that we have carried out a primary survey on a site, using a regularly sampled grid, and have classified the stands into different associations or vegetation types. Provided that the distances between the stands is not too great compared with the size of the stand, then a reasonably accurate vegetation map of the area can be constructed to show the distributions of the different vegetation types. Reasons for more detailed field work are varied, but many would involve examination of the quantity of a species in the different associations. Comparisons require statistical tests and so random sampling is preferable. However, it is not now appropriate to randomly sample over the whole area since we may find that some of the vegetation types have extremely few stands and vice versa. It is, however, perfectly legitimate to randomly sample within each vegetation type separately and so ensure equal samples in each type. Such a scheme is known as **stratified random sampling**.

17

Vegetation parameters

For accurate work, vegetation parameters are best estimated by actual measurement rather than by eye. Even if this is done we still say that the quantity is *estimated* rather than *measured*, since the latter signifies an accuracy which is normally unobtainable in the field. In this section the commonly used parameters will each be examined in turn, and in the next section aspects of their own peculiar sampling problems requiring attention will be dealt with.

Vegetation parameters are of two broad types – **absolute** and **non-absolute** – the latter depend on the precise sampling method used, whereas the former do not. Density, cover, biomass and performance are absolute; frequency is non-absolute.

DENSITY

Density is number in a given area. What unit is actually counted in the area depends on the species in question. Ideally the unit is an individual plant, and this is indeed applicable to **monocarpic** species (annuals and biennials) and also to trees and shrubs to a greater or lesser extent. Herbaceous **polycarpic** species (perennials) usually have a complex morphology due to continued vegetative growth of parts at or below ground level from year to year, and an individual may not be recognizable. If an estimate of density is required in these instances, any convenient morphological unit can be used. For example, in species having extensive underground rhizomes where aerial shoots develop at the nodes, each aerial shoot is a convenient unit; two such plants are *Mercurialis perennis* (dog's mercury) of woods and *Carex arenaria* (sand sedge) of sand dunes. Perennial grasses are notoriously difficult in the context of density, and a variety of morphological features have been used, depending on species. Thus, tufts or tussocks might be appropriate, as in *Brachypodium sylvaticum* (wood false-brome) and *Deschampsia caespitosa* (tussock grass), whereas in a species such as *Holcus mollis* which tends to produce individual tillers from an underground rhizome, giving the appearance of what Hubbard (1984) calls a 'loose mat', each aerial shoot or tiller could be counted. Strictly, in any grass species – annual or perennial – it could be argued that the individual tiller should be counted, as this is the basic space-occupying unit in all grasses. This is, of course, very laborious and normally impractical for all but the smallest scale work.

The field measurement of density is very simple in principle; the number of the selected unit is counted within the area of the quadrat. It is, however, very desirable that the number be finally expressed in relation to a standard area to facilitate comparison with other work, and in SI units the square metre (m^2) is appropriate. Of course, individuals of a plant species are not usually regularly or even randomly distributed, and if you are using, say, a square 0.25 m^2 quadrat, multiplication by four will give you the result on a m^2 basis. But if you had actually sampled 1 m^2 instead of 0.25 m^2 at the same location, it is most unlikely that your measured density would be anywhere near four times the figure you obtained with the latter size quadrat. This is due to the fact that nearly all plant species under most conditions occur in a clustered manner (see Kershaw & Looney 1985, for a full discussion on this topic). Hence, sampling with a quadrat whose area is not the same as the area in which density values are to be ultimately expressed does distort the picture. On the other hand for many purposes in most vegetation types, a stand of area 1 m^2 is not a suitable size: it is too large for herbaceous vegetation and too small for woody species.

COVER

The **cover** of a species is defined as the proportion of ground occupied by perpendicular projection on to it of the aerial parts of individuals of the species under consideration. The meaning behind this formal definition can be appreciated by thinking of an area of vegetation consisting of one species only on level ground; then, if this area were illuminated vertically from above, the proportion of the ground in shadow is the cover of the species (Greig-Smith 1983).

Cover is normally expressed as a percentage and, from the interpretation of the formal definition in the previous paragraph, it is obvious that the maximum cover of any one species is 100%. But in normal multi-species vegetation the total cover of all the species present is usually greater than 100%. If the total cover is less than 100% then the vegetation is 'open' with gaps in the foliage canopy; otherwise, the vegetation is 'closed' with no appreciable gaps in the canopy.

The principle of measuring cover, as opposed to estimation by eye, is by determining the presence or absence of any part of an individual of that species at a *point* on the surface of the ground. Employment of this principle is known as the **point quadrat** method, and strictly comes under the heading of plotless sampling methods. In practice, a large number of points are sampled and the percentage of these points at which the species occurs is, directly, its percentage cover.

The most accurate measurement of cover by the point quadrat method is obtained by the use of an optical cross-wire apparatus focused on the vegetation. Of course, this is excruciatingly laborious; but it has been shown that, whatever is used to define the point quadrat, the result obtained is profoundly affected by the size of the 'point'. The nearer the 'point' is to a true geometrical dimensionless point the more accurate the cover measurement.

In practice, a point is defined by lowering a pin vertically through the vegetation and noting the species that it touches as it descends. A pin needs to be held by some means other than the operator, and it is usual to hold several pins together in a frame (Plate 2). Thin knitting needles make useful pins, bearing in mind that the point should be as fine as possible. Grassland, especially where the turf is short and density measurements well nigh impossible, is a very common habitat where cover is used when a measured estimate of abundance is desired. Nevertheless, it requires patience and good conditions to make the measurements (Plate 3).

BIOMASS (YIELD)

Biomass of each species in vegetation is a parameter that seems to be gaining in popularity. It is measured by clipping the vegetation to ground level, sorting the clippings into species, and drying each in an oven to constant weight. Root excavations are very rarely carried out; it would take very much longer, and the amount of root material acquired would inevitably be incomplete. Not only that, but total biomass measurement ensures complete destruction of the vegetation in the stand, whereas taking shoot material results in only partial destruction: some species will regenerate, others will not.

PERFORMANCE

A variant of the yield parameter is **performance**. Common measures are leaf length or leaf width, or some combination of the two to represent leaf area – an indicator of the area available for photosynthesis. Other suitable morphological features are: flower number, length of flowering spike, or number of seeds per capsule – all measures of the reproductive performance of the plants.

FREQUENCY

The **frequency** of a species is defined as the probability of finding it within a quadrat when the quadrat is placed on the ground. A probability measure always lies between 0 (species never occurs) to 1 (species always occurs) and so frequency is usually expressed as a percentage.

Frequency is actually estimated in the field by placing a quadrat on the ground in many different places within the desired area, noting how many placings contain an individual of the species in question, and expressing this as a percentage of the total number of placings. To do this requires a relatively large area to place the quadrat in, whereas we may require frequency in only a small area. Here, a relatively large square quadrat (say

1 m^2) could be subdivided into 100 1 dm^2 squares with thin nylon cord. The quadrat is then placed on the ground, and the number of sub-quadrats containing the species in question is, directly, the frequency. The measure obtained in this way is known as **local frequency**.

We have used the phrase 'containing the species'; this is vague. A quadrat containing the species under consideration may merely mean that a small part of the foliage is present in the quadrat. So, we define two types of frequency:

(a) **shoot frequency** is obtained by considering a species 'present' when *any* part of an individual of that species occurs in the quadrat;
(b) **rooted frequency** only includes a species as 'present' when it is actually rooted within the area of the quadrat.

Species will, of course, give higher shoot than rooted frequencies, and for the following discussion we mainly assume rooted frequency.

As already mentioned, frequency is a non-absolute measure, which means that the actual result obtained is affected by three apparently irrelevant factors – quadrat size, individual plant size (mainly shoot frequency) and the spatial distribution of individual plants. As regards quadrat size, consider the situation shown in Figure 2.3. If the large area containing 'plants' was sampled by quadrat A, 100% frequency would be obtained; whereas with quadrat B the result would be lower. Thus, quadrat size should always be quoted with a frequency estimate, and the same quadrat size should be adhered to when surveying different areas for comparative purposes.

The plant size factor is mainly relevant to shoot frequency, and is illustrated in Figure 2.4. The two plant species A and B are of equal density,

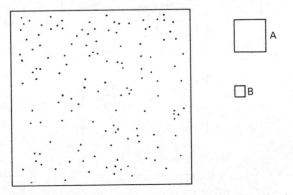

Figure 2.3 The dependence of frequency on quadrat size (after Kershaw & Looney 1985, by permission of Edward Arnold). For explanation, see text.

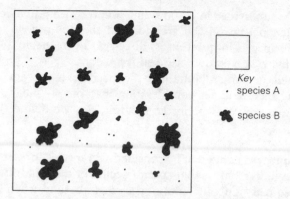

Figure 2.4 The dependence of frequency on plant size (after Kershaw & Looney 1985, by permission of Edward Arnold). For explanation, see text.

and they are sampled by the quadrat shown. Species B would give 100% frequency, but species A would show a lower value.

The spatial arrangement of the individuals of the species is a more complex phenomenon. It has been shown on innumerable occasions that individuals of a plant species are rarely, if ever, randomly arranged, so we are left with either a regular pattern (rare) or clusterings of different degrees. Figure 2.5 shows the situations and a sampling quadrat. The three areas are all equal, and so is the number of 'plants' in each; hence, the overall density is the same in each area. However, a frequency estimate in area A would be 100%, in area B nearly zero, and some intermediate value in area C.

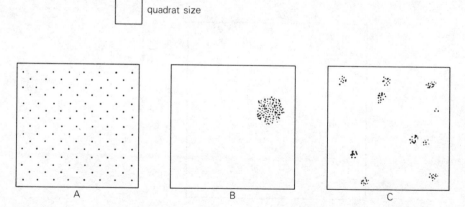

Figure 2.5 The dependence of frequency on spatial pattern (after Kershaw & Looney 1985, by permission of Edward Arnold). For explanation, see text.

In the discussion above on the three artificial situations, frequency estimates have mostly been compared with the density of the 'plants'. This is natural because density is the basis on which these situations have been created, the exception being the plant size example. If the individual plants occur on the ground at random, or if the clustering can be defined mathematically, then a formula relating frequency and density for a quadrat of *given size* can be derived. However, since plants scarcely ever occur at random on the ground, or in some mathematically definable clustering pattern, such formulae would have almost no practical use. Relating other absolute parameters to frequency is theoretically possible, cover should be particularly easy, but no one has ever done this as far as I am aware.

Special sampling problems

The following discussion is intended to supplement the material presented above, dealing with sampling problems peculiar to each of the vegetation parameters.

DENSITY

Three decisions have to be made before the field work is undertaken: (a) the size of the sampling quadrat, (b) the shape of the quadrat and (c) the number of samples to be taken. These are not independent decisions in practice.

With regard to quadrat size, we first note that if the individuals of the species under study are randomly distributed on the ground, then the accuracy of the density estimate is only dependent on the number of individuals counted. This is because, under this condition of random spatial pattern, the numbers of individuals occurring in randomly placed quadrats follows the Poisson distribution, in which the variance is equal to the mean. Let x be the total number of individuals counted in n stands; then both the mean number per quadrat and the variance of a single quadrat is x/n. Thus, the variance of the mean of n quadrats is x/n^2, and so the standard error of the mean is $\sqrt{x}/n$. Finally, we see that the ratio of standard error of the mean to the mean itself is

$$\frac{\sqrt{x}}{n} \cdot \frac{n}{x} = \frac{1}{\sqrt{x}}$$

so the precision of the estimate of the mean is directly proportional to the square root of the total number of individuals counted.

In the more usual case of non-randomness it is best to use a small quadrat, but not so small that it approaches the size of the individual plants.

23

A small quadrat makes it easier to count the number of individuals therein, but on the other hand there is an increased **edge effect**. The difficulty about an edge is deciding exactly what is within and what is outside the quadrat, so the larger the ratio of length of edge to area of quadrat, the 'more' edge there is to a quadrat and so the greater the edge effect. Note that with a square quadrat of side l, the ratio of edge to enclosed area is given by $4l/l^2 = 4/l$. Hence, the smaller the quadrat, the larger the ratio and so the greater are the problems of the edge effect.

With regard to quadrat shape, it has long been customary to use a square quadrat, and not much thought has been given to using other shapes. What little work has been done on this question has tended to show that the precision of the estimates of mean density are greater when a rectangle is used than when a square is employed (Greig-Smith 1983). For a rectangle where the length of the longer side is n times that of the shorter side, b, the ratio of sides to enclosed area is

$$\frac{2(nb + b)}{nb^2} = \frac{2(n + 1)}{nb}$$

which varies from $4/b$ when $n = 1$ (a square, see above) down to a limit of $2/b$ as n becomes indefinitely large (a very thin rectangle). For a rectangle twice as long as broad the ratio is $3/b$, and this ratio is probably the optimum in practical terms.

In the absence of a random distribution of individual plants on the ground, the number of sample quadrats required for a certain degree of precision of the mean density estimate cannot be determined unless the variance of density from one quadrat to another is known. This would necessitate doing a pilot study first. Alternatively, one could proceed with the main sampling, and plot a graph of mean density per quadrat against number of quadrats, say, every ten quadrats. The mean density will probably oscillate greatly for low quadrat numbers; but as the sample size increases, the mean density estimates will tend to stabilize. From such a graph, one could visually judge when a sufficient number of quadrats have been used.

BIOMASS

Biomass provides fewer sampling problems than the other vegetation parameters. The only problem of note, as with any kind of sampling, is that of sample size. The approach suggested for density is not feasible, because biomass estimates involve laboratory work of sorting into species, drying and weighing; whereas density estimates can be calculated and graphed straight away in the field. Only a pilot study, to gain an idea of the variability involved, can be suggested. Remember, however, that biomass sampling involves at least the partial destruction of the quadrat area; sample numbers should be the minimum possible, and there must be good reasons for using bioimass as a vegetation parameter.

FREQUENCY

The prime requirement in estimating frequency is to use as large a sample size as possible. Since a frequency measure is easy to make (it only requires finding one individual in a laid quadrat), this will not be too great a hardship. The reason for requiring an especially large sample size will become apparent in the next section on statistical tests.

COVER

Sampling of percentage cover is very similar in principle to the sampling of percentage frequency; we are, in effect, recording frequency with a very small, ideally infinitesimal, quadrat. Thus the recommendation of having a large sample size is also applicable to percentage cover. However, there are two aspects of the sampling of percentage cover with point quadrats that are not applicable to the sampling of frequency with ordinary quadrats. These aspects are: (a) the effect of pin diameter and (b) the effect of using a frame containing several pins rather than just one pin.

The effect of pin diameter on the estimate of cover obtained has already been referred to (p. 20). Ideally a point quadrat should be indeed a dimensionless point, but this is virtually unobtainable in practice except with an optical cross-wire in a sighting apparatus which is scarcely a practical tool for making measured cover estimates in the field. The true cover will be over-estimated by using a large diameter pin, but the error will not be the same between species. The apparent increase in cover, associated with a large diameter pin, will be greater in species having small, or elongated, or dissected leaves, than in species with large, undissected leaves. Clearly, the tip of the pin must be as small as possible.

Very commonly, frames containing ten pins are used. However, with one placement of the frame and the lowering of the ten pins, the hits by each pin are not independent of one another unless the plants are spatially random, or are in very small clusters compared with the inter-pin distances. This non-independence of the pins has the effect of increasing the variance of estimate of the cover, i.e. lowering the precision of the estimate. This can be shown theoretically; and Goodall (1952) has also demonstrated the decrease in precision, using a frame of ten pins compared with a frame of one pin, in the field. This implies that the precision of cover estimate if a one-pin frame is placed at random 100 times is greater than if a ten-pin frame is placed at random ten times, even though in each case 100 pins are lowered into the vegetation. However, in the former case, 100 random points have to be located rather than ten.

25

Statistical analyses

Although not strictly 'field methods', the statistical treatment of vegetation parameter sample data is conveniently dealt with here.

The main requirement for statistical analyses of such field data is the comparison of parameter estimates of a species between different areas or, more rarely, the comparison of parameter estimates of different species in the same area. The latter activity, however, may make for interpretational difficulties; for example, comparing the densities of two species of very different size, such as *Circaea lutetiana* (enchanter's nightshade) shoots with those of *Polytrichum commune*. Hence, in this section, we shall have in mind the comparison of parameter estimates of a single species between different areas.

DENSITY

If the spatial arrangement of individual plants of the species of interest were random, then the number of individuals per quadrat (i.e. 0, 1, 2, ...) would tend to follow a Poisson distribution. The Poisson distribution has only one parameter, which is both the mean and variance of the distribution; consequently, these quantities are not independent as they are in a normal distribution. A Poisson distribution having a large mean has a high variance also, and vice versa. Further, a Poisson distribution with a low mean is highly positively skew, but the skewness decreases as the mean increases. A Poisson distribution with a mean of 10 or more has a probability density close to the normal in form (apart from the fact that a variate having a Poisson distribution can only take integer values, whereas a variate which is normally distributed can have any real value).

Now the usual statistical methods for comparing sample means – Student's-t test for two samples and analysis of variance (ANOVA) for more than two samples – require the data to be approximately normally distributed, and for the group sample variances to be essentially homogeneous. Now if the group means of density measurements are sufficiently high, the normality requirement is met but the variance homogeneity condition is not. However, if the square root transformation is applied to the data before analysis (either t test or ANOVA), the sample variances tend to become independent of the sample means, and so the data are valid for analysis. The square root transformation implies simply taking the square root of every density value in the data set, i.e. $\sqrt{x}$, where x is a density. If any sample mean in the whole data set is small, say less than 10, then the transformation $\sqrt{(x + 0.5)}$ or $\sqrt{(x + 0.375)}$ is better.

Very rarely, however, are the individual plants randomly distributed on the ground; usually they occur in clusters, and this will give rise to sample variances greater than sample means. Nevertheless, variances of sample

26

density data still tend to be *proportional* to the means, and so the square root transformation should be regarded as a standard procedure in analyzing density data.

BIOMASS

Since plant growth is essentially a multiplicative, rather than an additive, process, biomass estimates tend to be lognormally distributed. Such data should be logarithmically transformed before analysis: each observation, x, is converted to $\log_e x$.

FREQUENCY AND COVER

As previously mentioned, frequency and cover are similar quantities from a statistical viewpoint. For each placing of a quadrat (ordinary or point), the species of interest either does or does not occur; hence, the proportional (or percentage) frequency or cover estimates are binomially distributed, and this applies regardless of the spatial distribution of the plants so long as quadrat placement is random. In this connection, note that in the positioning at random of a frame of, say, ten pins, only *one* of these pins is actually randomly placed; the remaining nine are fixed in relation to the one pin by the design of the frame.

The mean of a binomial distribution is np, where n is the sample size and p is the proportion of occurrences (i.e. the frequency or cover expressed as a proportion rather than as a percentage); the variance of the distribution is $np(1 - p)$, so, as in the case of the Poisson distribution, the mean and the variance of the binomial distribution are not independent. If p departs markedly from 0.5, the binomial distribution becomes very asymmetric unless n is large. Hence, a t test or ANOVA should not be employed for comparing frequency or cover estimates of a species from different areas; the contingency table method, as demonstrated in the example below, should be used.

Example 2.1

In two areas, A and B, the frequency of a species was estimated to be 51% and 62%, respectively, on the basis of 100 random quadrats in each area. Do these results indicate a real difference in frequency of this species between the two areas?

Since the frequency estimate is 51% in area A, this implies that the species was present in 51 and absent in 49 of the 100 quadrats. In area B, the species occurred in 62 and did not occur in 38 of the quadrats; so the contingency table appears thus:

	Species +	Species −	
Area A	51	49	100
Area B	62	38	100
	113	87	200

This 2×2 contingency table has the same structure as that in Table 5.1, although the nature of the rows differs between the two cases. The method of calculation is shown on page 75, and for the present data is

$$\chi^2_{[1]} = \frac{200\{(51)(38) - (62)(49)\}^2}{(113)(87)(100)(100)} = 2.46$$

For significance at $P(0.05)$, with 1 degree of freedom, χ^2 needs to be 3.84; hence, our calculated value of χ^2 is not significant, and so there is no evidence of a real difference between the frequencies of this species in areas A and B.

Plotless sampling

In forest or scrub vegetation the establishment of sample stands, large enough to ensure that each contains an adequate number of individual plants for a realistic estimate of a vegetation parameter, becomes a major practical difficulty. Obviously, areas of many square metres are required and must be delimited by pegs and string. While this procedure may not be too difficult in a wood containing little or no undergrowth, it would become impossible in areas where the field layer was dense, particularly if this layer was dominated by brambles (*Rubus fruticosus*) or similar thorny species. The concept of plotless sampling overcomes many of the difficulties encountered in vegetation dominated by woody species and, indeed, may be a useful method to apply in herbaceous vegetation in certain circumstances. Although primarily a set of methods pertaining to more detailed fieldwork, the results obtained by the use of plotless sampling techniques may also be of the kind required in primary survey.

There are several methods of plotless sampling, but all are based on the idea of measuring distances from randomly chosen points in the study area to certain individual plants, usually the nearest. The species names of the individual plants, whose distances from the random points are measured, must also be recorded and other features of the plants may be noted. In particular, the basal diameter of tree trunks will enable a size distribution to be plotted.

Point-centred quarter method

Probably the best of the plotless sampling techniques is the **point-centred quarter method**. The basis of the method is shown in Figure 2.6. A point is established at random in the study area, four quadrants around the point are marked, the distances from the point to the nearest tree in each

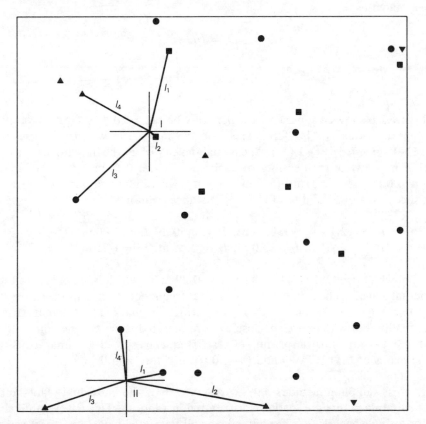

Figure 2.6 The point-centred quarter method, applied to woodland trees. The + represents the selected random point, around which four quadrants are established, and the distance to the nearest tree, $l_i (i = 1, 2, 3$ or $4)$, is measured in each quadrant. The symbols •, ■, ▲, ▼ represent four different tree species.

29

quadrant are measured, l_i ($i = 1$, 2, 3 or 4), and the species name is also recorded in each case. The procedure is repeated for many random points; then the density of each species is estimated as follows.

First, the density of all the trees, regardless of species, in the study area is estimated. This is done by averaging all n length measurements, $\bar{l} = (\Sigma\, l)/n$. Then overall tree density is estimated as:

$$D = 1/(\bar{l})^2$$

Next, the frequency of each tree species encountered is obtained as a proportion of the total number of distance measurements made: $f_j = n_j/n$, $j = 1, \ldots s$ where s is the total number of species, and n_j is the number of recorded distances to species j. Finally, the density of species j, D_j, is estimated as:

$$D_j = f_j D$$

Example 2.2

Figure 2.6 shows a square area which may be regarded as 100 m^2. Twenty-eight individuals of 4 different species occur in the area with frequencies of 14 ($\bullet$), 7 ($\blacksquare$), 5 ($\blacktriangle$), 2 ($\blacktriangledown$). The actual density of all the individuals is thus 0.28 m^{-2}, while the densities of each species are: 0.14 m^{-2} ($\bullet$), 0.07 m^{-2} ($\blacksquare$), 0.05 m^{-2} ($\blacktriangle$) and 0.02 m^{-2} ($\blacktriangledown$). Two random points were established, as shown, and the following distances obtained:

Point I: $l_1 = 2.15$ m, $l_2 = 0.15$ m, $l_3 = 2.50$ m, $l_4 = 2.00$ m;
Point II: $l_1 = 0.90$ m, $l_2 = 3.60$ m, $l_3 = 2.25$ m, $l_4 = 1.25$ m.

from which $\bar{l} = 1.85$ m, $D = 1/(1.85)^2 = 0.29$ m^{-2}, which is very close to the actual density. The *observed* proportional frequency of each species is 0.375 ($\bullet$) and ($\blacktriangle$), 0.25 ($\blacksquare$) and 0 ($\blacktriangledown$), which is markedly different from actuality; but with only eight observations at two distinct sample points, the discrepancy is not surprising. These frequencies lead to final density estimates of 0.11 m^{-2} ($\bullet$) and ($\blacktriangle$), 0.07 m^{-2} ($\blacksquare$) and 0 ($\blacktriangledown$).

Plotless sampling methods only work properly if the individual plants and the sample points are spatially distributed at random. This is very unlikely for a single species, but more likely for all the individuals (e.g. trees) in an area, regardless of species. This is the reason why the density of all individuals must be obtained first, and then the density estimate of each species calculated in proportion to their frequency.

30

Although ideally suited to vegetation dominated by woody species, plotless sampling methods can be used in herbaceous vegetation (Plate 4). This may be useful in situations where the density of some species is high and counting within quadrats is tedious and possibly inaccurate. Measuring short point to point distances is much easier. A word of warning is required, however. Apart from the randomness requirements, described above, unbiased density estimates also require the distance measurements to be made from the random point to the *centre* of the individual plant. While this is impossible for trees or shrubs, the error involved by measuring to the trunk circumference is small. However, with (particularly) a rosette species of herbaceous plant considerable error would be incurred by measuring to the nearest edge of a leaf; the measured distance would be shorter than it should be, yielding a density estimate biased upwards. Fortunately, for plant species of this kind there is no difficulty in measuring to the centre.

3

Fundamental principles of analytical methods

In this chapter we shall first examine the conceptual framework upon which analytical methods are based. Secondly, the two broad categories of methods – classification and ordination – are introduced, and discussed in relation to vegetation structure and the analytical framework. Thirdly, we return to the distinction between qualitative and quantitative data, and consider these data types on the basis of the analytical framework and ecological considerations; and finally, we discuss the desirability of including or excluding species of very low occurrence in the data set.

The geometric model

In order to understand the workings of vegetation analysis methods, we need a graphical illustration of the mathematical representation of species-

Table 3.1 A set of artificial vegetation data (quantitative) for illustrating the geometric model.

	Species		
Stand	A	B	C
1	7	0	2
2	7	1	1
3	8	1	2
4	1	1	7
5	0	3	5
6	2	5	3
7	3	3	6
8	1	7	2

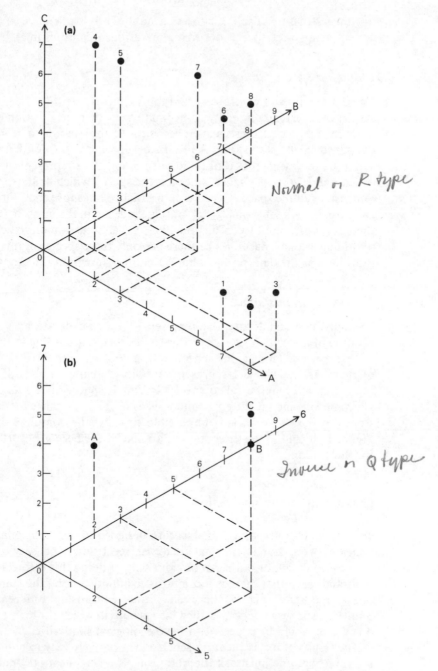

Normal or R type

Inverse or Q type

Figure 3.1 The fundamental geometric model for classification and ordination methods: (a) stands in species space; (b) species in stands space.

in-stands structure. Table 3.1 shows a small set of artificial vegetation data consisting of three species occurring in various quantities in eight stands.

Stands in species space

There are two ways of drawing a graph of these data (Fig. 3.1). First, the species may be represented by co-ordinate axes; then the position of a stand is given, relative to these axes, according to the amount of each species contained in the stand (Fig. 3.1a). In other words, a row of numbers in Table 3.1 represents the co-ordinates of a stand. This representation may be called **stands in species space**, since it is the species which define the different dimensions in the space. Of course, in practice not more than three species can actually be shown on a graph of this kind; but in the mathematical sense there is no such limit (see Causton 1983, for a discussion on the concept of multidimensional space). Since there are only three species in this example, the whole of the data set in Table 3.1 can be shown on the graph.

Species in stands space

Secondly, the stands may be represented by co-ordinate axes; then the position of a species is given, relative to these axes, according to the amount of that species contained in each stand (Fig. 3.1b). In other words, a column of numbers in Table 3.1 represents the co-ordinates of the species. This representation may be called **species in stands space**, since it is the stands which now define the different dimensions in space. Because there are eight stands in the example, it is impossible to show the complete data set in Figure 3.1b, and so the three stands 5, 6 and 7 have been arbitrarily selected for illustration.

Classification and ordination

The results of a vegetation analysis are designed to highlight relationships: either between stands (normal analysis, see below), or between species (inverse analysis), or both. An obvious way of doing this is to classify, say, the stands such that there is more floristic affinity among the stands of any one group than there is between stands comprising different groups. Similarly, species may be classified into groups in which the members of any one group would be expected to have ecological similarities.

The results of a classificatory scheme are relatively easy to assimilate and interpret, but one must ask the question, 'is a classification "natural" for the vegetation under scrutiny, or are groupings being imposed by the method on vegetation in the study area which actually varies smoothly from

34

point to point?' This brings us right back to the community/continuum ideas on the nature of vegetation discussed in Chapter 1; at first sight it would seem that classification is appropriate for vegetation having a distinct community structure, but not for a continuum. In terms of our geometric model, described in the previous section, stands taken from different communities would show a clustering effect – much as they do in the data set of Table 3.1. Thus, stands 1, 2 and 3 show as a distinct cluster; stands 4 and 5 are relatively close together, and stands 6, 7 and 8 form a looser grouping (Fig. 3.1a). If, on the other hand, sample stands from a continuum were plotted as in Figure 3.1a, there would be no distinct clusters but simply a more or less structureless swarm of points in the space.

The other broad category of vegetation analyses is known as ordination, and the results for, say, a stand ordination consist of a set of co-ordinate values for each stand which can be plotted on one or more graphs. Ordinations are ideally suited to the continuum situation. It might also be thought that they would be suitable for vegetation showing a community structure, by showing up the communities as clusters of points; but currently available ordination methods are rather poor in this respect.

At this point then, we are left thinking that we classify vegetation if it consists of a series of communities, and ordinate it if changes are smooth rather than discontinuous. Since our discussion in Chapter 1 showed that it may be difficult to be sure what an actual situation is – it rather depends on the scale at which we are working – we may thus be unsure whether to classify or ordinate our vegetation data. Our decision might be simplified if we can equally well ordinate community structured vegetation and classify a continuum.

There is, in fact, nothing to prevent us applying ordination methods to vegetation which changes discontinuously so long as we do not expect a clear clustering effect to be revealed on the resultant graphs. On the other hand, classification of a continuum implies superimposing a community structure, which is an artifact. If, however, a prime aim of a vegetation analysis is simplification, then undoubtedly a classification is easier to assimilate than an ordination. Now, one can always classify a number of objects into groups, and there are many methods or classification strategies for doing this. The objects here are stands, and if the vegetation in a given study area really did comprise distinct communities, then all good classification strategies should yield essentially the same results. However, different classificatory methods applied to a continuum will tend to yield different results. It thus becomes vital to examine the detailed workings of classification methods employed, otherwise it will not be possible to interpret the results intelligently.

With the above points in mind, it is not so much the structure of the vegetation studied that will determine the analytical methods to be employed, rather it is the purposes of the study that guide us to appropriate analytical techniques. Indeed, many investigators carry out a range of analyses on their data in order to discover different facets about the vegetation studied, as different analyses highlight different aspects.

Normal and inverse analyses

Table 3.1 shows that there are two types of variable in vegetation data: species of plant and sample stands of vegetation. Vegetation analyses enable us to classify and ordinate both these categories. Classification of stands is often referred to as **normal** classification, whereas the classification of species may be called **inverse** classification. The same terminology could be extended to ordination, but here it is usual to speak more directly of a stand ordination or a species ordination. There is an older terminology which does have certain uses (see Ch. 7): a stand ordination is known as an **R-type** analysis and a species ordination is called a **Q-type** analysis. The geometric basis for stand classifications and ordinations (normal or R-type) is that depicted in Figure 3.1a, while that for species classifications and ordinations (inverse or Q-type) is shown in Figure 3.1b.

Qualitative and quantitative data

We now return to the subject of vegetation data type, which has been under scrutiny in each of the first two chapters. In the first chapter the nature of these two kinds of data was described, and in the second chapter some recommendations were made to assist in the decision regarding which kind of data to acquire in a vegetation study. We are now in a position to gain insight into the difference between qualitative and quantitative data in relation to analytical methods, through consideration of the geometric model. Also, we shall discuss qualitative versus quantitative data in relation to ecological criteria.

Mathematical considerations

In the geometric model as exemplified by Table 3.1 and Figure 3.1 the data are clearly quantitative. For qualitative data the entries in Table 3.1 would consist solely of zeros and ones. An example of such data is given in Table 3.2. When stands are plotted in species space (Fig. 3.2a), because there are three species (and hence three dimensions) the stands are arranged at the vertices of a cube of unit side. The data in Table 3.2 have been specially constructed such that each of the eight stands is different from any

Table 3.2 A set of qualitative artificial vegetation data for illustrating the geometric model.

	Species		
Stand	A	B	C
1	1	1	0
2	1	0	1
3	0	0	0
4	0	0	1
5	1	1	1
6	0	1	1
7	0	1	0
8	1	0	0

of the others, and so one stand occurs at each of the eight vertices of the cube. If there are more stands than one having the same species make-up, then two or more dots are superimposed at the appropriate vertex. In Figure 3.2a, stands 2 and 7 only *appear* to be superimposed because of the orientation of the axes. Figure 3.2b shows the plot of species in stand space for three stands. As there are only three species, and each has a different pattern of occurrence within the three stands (5, 6 and 7), then only three of the eight vertices of the cube are occupied by a species.

For the remainder of this discussion we shall concentrate on the aspect of stands in species space. On comparing Figure 3.2a with Figure 3.1a, it might be considered that there was little information in the former compared with the latter. This first impression is reinforced if one considers the situation obtaining with just one or two species (Figs 3.3a & b); the number of distinct possibilities of the stands with respect to the species axes appears decidedly limited. However, if the sequence from one to three species is examined in order (Fig. 3.3a, Fig. 3.3b, then Fig. 3.2a), it can be seen that the number of possible positions increases with species number: two positions for one species, four for two species, and eight for three species. In fact, for m species the number of distinct points in species space which stands can occupy is 2^m; so with the usual number of species occurring in real data — at least double figures — it can be appreciated that there is plenty of 'freedom' for informative patterns of stands in species space to be shown. Even with only ten species, which is very minimal for a vegetation survey in almost any type of habitat, there are $2^{10} = 1024$ vertices of the ten-dimensional hypercube and so 1024 distinct positions for stands

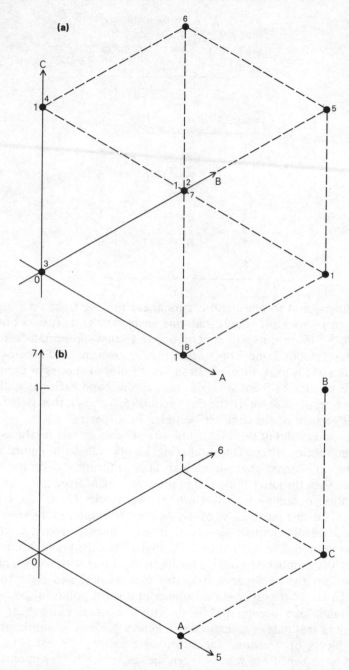

Figure 3.2 Qualitative data geometric model: (a) stands in species space; (b) species in stands space.

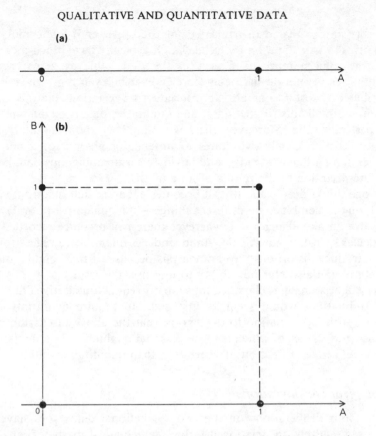

Figure 3.3 Stands in species space: (a) one species; (b) two species.

to occur. Of course, even in such a small species number data set, perhaps having only 100 stands in the sample, there might be many stands having the same species composition (superimposed points at some vertices); but the important fact is that there is *potential* for considerable variation in the pattern of stands in species space in qualitative vegetation data.

Another feature of species-in-stands data is that they contain an excessive number of zeros: the artificial data of Tables 3.1 and 3.2 are quite atypical in this respect. What exactly does this statement mean? Let us for a moment imagine that the columns of Table 3.1 represent not quantities of three different species, but levels of environmental factors; say, soil pH, percentage soil moisture at field capacity, and a light intensity value expressed as a percentage of full daylight. If this were the case, no figure in the data table would be zero. Not only that, but if a sufficient number of stands had been sampled (rows in the data table) it could well be found that soil pH approximated to a normal distribution and that soil moisture and light intensity approximated to truncated (at the upper end) lognormal distributions. Floristic data, on the other hand, typically have much more than half the total number of figures in the data table as zeros; so even quantitative vegetation data can scarcely ever be even remotely normally distributed. Most techniques appropriate to vegetation analysis do not require normally distributed data, and qualitative data (noughts and ones) are perfectly valid. Moreover, there is a simplicity about qualitative data because there are only two states – presence or absence of a particular species in a particular stand – and having clear unambiguous data helps in the interpretation of the results of an analysis.

If one thinks deeply about it, it becomes apparent that quantitative data are trying to combine two different things – the qualitative (0 or 1) aspect and also an abundance value wherever there is a presence recorded for a particular species in a particular stand. Indeed, quantitative vegetation data can introduce distortions into an analysis because of their highly anormal structure, and also blur our ability to interpret the results.

As will be seen later, classifications mostly require qualitative data (Ch.6) and ordinations would seem at first sight to require quantitative data (Ch.7); so it is in relation to ordination that the above discussion chiefly applies, the essence of which is to suggest that qualitative data may provide a more effective ordination of vegetation than quantitative data.

Ecological considerations

So far, our deliberations on the two vegetational data types have been discussed entirely in relation to the mathematical basis of analytical methods. You may be left wondering whether this is the be-all and end-all of the subject, and that biological and ecological considerations find no place. This is not true, and indeed should not be true in what is, after all, a field activity in which the aim is to further a particular branch of biology – plant ecology.

When a sample stand is delimited, two questions are asked in an indirect way. Does a particular species occur in the stand or not (qualitative), and if so, how much of that species occurs in the stand (quantitative)? Assuming that propagules of the species in question are not limiting, two factors govern the occurrence of a species. First, the physical factors of the environment of the sample stand must be within the tolerance range for seed germination and seedling growth (tolerance ranges are usually narrowest in the very young seedling) in order that a plant may become established; and secondly, the species must have sufficient competitive ability *in the environment of the sample stand* in order to survive. It can be appreciated that if the environment of the sample stand is within the tolerance range of seedlings of the species then, still assuming that propagules are available, one or more plants may become established. The actual number that establish and survive will be partly affected by the number of propagules available, and partly by a host of other factors which are conveniently lumped together as 'chance'. Thus, particularly if one is concerned with ecological purposes of vegetation analysis, from the viewpoint of species tolerance the question of importance is whether or not a species actually occurs at all at a particular place, rather than how much of it is there.

The second broad factor governing the occurrence of a plant species in a particular place − competitive ability or aggression − however, will affect both occurrence and amount. Obviously the species in the environment of the stand must be sufficiently aggressive to survive at all, but differing degrees of competitive ability must to some extent affect the amount (in terms of number of individuals, or size of individuals, or both) of the species present. Putting these two factors − tolerance and aggression − together, we see that occurrence or non-occurrence of a particular species at a particular place on the ground is the crucial point, with the abundance aspect being rather secondary.

In his book, Grime (1979) defines three categories of plant: competitors, stress tolerators, and ruderals. Only in the first group is the aggression aspect particularly important and so, viewing all plant species together, the relative importance of qualitative as against quantitative data is weighted even more in favour of the former.

Another difficulty with quantitative data in a primary vegetation survey involving many species is the fact that species which vary so greatly in their size and morphology are being compared. For example, it is really meaningless to compare densities of, say, hogweed (*Heracleum sphondylium*) with the moss *Polytrichum formosum*. Similarly, it is equally useless to compare the cover of, say, *Circaea lutetiana* with that of *Anemone nemorosa* (wood anemone). In spring the shoots of the former species are only just beginning to emerge, while those of the latter are at full development; and in summer *Circaea lutetiana* has its maximum cover while the aerial parts of *Anemone nemorosa* have died back.

You should, however, clearly understand that all the above discussion is in relation to the primary survey situation. The geometric model and the ecological aspects considered all relate to the methods and mode of thought applicable to primary survey involving many species. When carrying out detailed work on one or a few species, abundance is of much greater interest, and the analytical methods are different. The essence of all the above is that while quantitative vegetation data can provide useful additional information, and should be acquired even in a primary survey, it should be used with caution in those analytical procedures which are used for primary survey data, i.e. the methods described in Chapters 6 and 7.

Species of low occurrence in a data set — retain or discard?

The final section of this chapter is concerned with a problem which, at first sight, would be thought trivial. In any set of vegetation data the species present will cover a wide span of frequencies. Only extremely rarely will any species occur in every stand but, at the other end of the scale, there will always be a number of species present which occur in only one stand of the data set. The question to discuss now is, should species with a low occurrence *in the data set* (one stand only, or perhaps two or even three or more, depending on the total number of stands in the data set) be eliminated before analysis? The italics in the above sentence are important: we are not concerned here with rare species as such, but rare in our particular set of data.

The importance of low frequency species lies in the fact that many of the analytical techniques are very sensitive to small changes in the data, such as the elimination of even rare species. If, therefore, one could end up with two rather different analytical results from essentially the same data (the only difference being the elimination of one or a few species of low occurrence), then one must ensure that the data submitted to analysis are as good as possible. Why might it be necessary to do what many would regard as an objectionable practice — data 'fiddling'? The answer to this question hinges on what we regard the status of these rare species to be.

Let us consider a species which has occurred in only one of the sample stands in our data set. Is the species there purely by chance, or is it there because it is a normal component of that vegetation type? The idea of a species being characteristic of a particular vegetation type is very much a community concept; but even in vegetation which is an obvious continuum, certain species tend to occur together, and we can call such groupings **species assemblages**. This is a more neutral term than a 'community of species'. If a species is present in a stand where ecological experience suggests that it is anomalous, then the species contributes little or no ecological information and should be discarded from the data set because it acts as unwanted background 'noise' and may unnecessarily distort the analysis. If the species is not in the sample stand simply by chance, then even an isolated occurrence contributes ecological information and the species should not then be eliminated from the data set prior to analysis.

It is impossible to generalize on the conditions to be taken into account in order to arrive at a decision on this question. The status of the species in question in the habitat under study must be carefully assessed in the light of its known ecology and the environment in which the species is growing. Historical evidence is very relevant here. If it is known that the vegetation on the site has changed markedly from that which is evident from old records, for example the incursion of scrub and woodland onto grassland, then the single occurrence of an anomalous species might be a relic of previous vegetation. Clearly, each case must be taken on its merits. The only general piece of advice that can be given is, 'if in doubt, don't': data should not be tampered with without good cause.

4

Case studies – introduction

The case studies are introduced here, before any analytical methods have been treated in detail, so that the reader will be in a position to fully appreciate the application of each method, as it is presented in Chapters 6–8, to real data sets. The analyses are collated in Chapter 9, but after reading about each method in Chapters 6–8 you may refer to appropriate applications in Chapter 9.

Since one of the aims of this book is to give guidance on interpreting the results of vegetation analysis, the case studies have been selected from those with which I have been more or less closely associated. The studies, which were carried out in England and Wales, comprise data acquired by students on a field course on the one hand and a research project on the other. The types of habitat and geographical distribution are thus quite narrow: the examples have been chosen first to illustrate what can be done using different field techniques in vegetation types of different complexity, and secondly to show the approaches used in interpretation.

Artificial Data

The first set of data presented is not really that of a case study at all; it is simply artificial data comprising ten stands and five species, without any underlying structure, which will be used to numerically illustrate most of the classification and ordination methods described in Chapters 6 and 7. The prime purpose of these species-in-stands data is to provide a small and manageable set whose main features can be readily appreciated. The Artificial Data are given in Table 4.1 in two forms: first, in the upper part of the table, in qualitative form where presence of a particular species in a particular stand is given the value 1, and the absence of a particular species in a particular stand has the value 0; and secondly, in the lower part of the table, in quantitative form where absences are still denoted by zeros, but presences are given an abundance value. These abundance values are not

Table 4.1 The Artificial Data, in both qualitative and quantitative form.

Stand	Species					Stand totals
	I	II	III	IV	V	
1	1	0	1	0	0	2
2	1	1	0	0	0	2
3	0	0	1	0	1	2
4	1	1	0	1	0	3
5	0	1	0	1	0	2
6	1	1	1	0	0	3
7	0	0	0	1	0	1
8	0	0	1	0	0	1
9	1	1	0	0	0	2
10	0	1	1	0	0	2
Species totals	5	6	5	3	1	
1	3	0	9	0	0	12
2	1	10	0	0	0	11
3	0	0	10	0	1	11
4	1	2	0	2	0	5
5	0	7	0	3	0	10
6	2	6	1	0	0	9
7	0	0	0	9	0	9
8	0	0	9	0	0	9
9	5	8	0	0	0	13
10	0	9	6	0	0	15
Species totals	12	42	35	14	1	

meant to represent any one kind of vegetation parameter, and they can take any integer (whole number) value from 1 to 10.

Iping Common – a lowland heath

Iping Common is one of several areas of heathland lying mainly on the Folkestone Beds of the Lower Greensand in the south-west corner of the Weald in West Sussex. Aspects of the ecology of Iping Common have been described by Harrison (1970).

After many years of non-management, with the consequent attainment of very extensive mature stands of *Calluna vulgaris*, most of the area suffered a severe uncontrolled burn in the dry summer of 1976. The severity of the fire was such that nearly all of the *Calluna vulgaris* regeneration was from seed rather than from the bases of old plants, and similarly for *Betula* spp. This means that in 1984, when the present study was made, few plants of any species (except *Pteridium aquilinum*, whose deep underground rhizomes were unaffected by the fire) were more than eight years old. This kind of historical fact is of vital significance in the interpretation of the existing vegetation.

Methods

FIELD

The data presented here were acquired by Honours Botany and Environmental Biology students on a field course in September 1984. A single line transect was laid down over a small part of the common, deliberately chosen to traverse several distinct vegetation types which were clearly visible to the eye. The transect was 120 m long, starting from an arbitrary point on

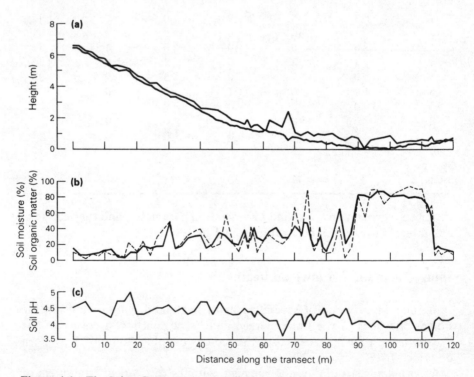

Figure 4.1 The Iping Common transect line: (a) height of ground above the lowest point (lower line), and height of vegetation (distance between the upper and lower lines); (b) percentage soil moisture (solid line) and percentage soil organic matter (dashed line); (c) soil pH.

relatively high ground and running due south down to and across a very shallow valley; the total height difference is little more than 6 m (Fig. 4.1a). The approximate National Grid Reference is SU850218. The transect was sampled every 2 m with square quadrats of area 0.25 m², but at some transition zones between different vegetation types sampling was done at each 1 m along the transect. The total number of sample stands thus delimited was 70.

Within every stand, the cover of each species present was estimated by eye, and a sample of soil from approximately the top 5 cm was taken. The height of each stand above the lowest point on the transect was ascertained by dumpy level and staff, and an estimate of the general vegetation height of the stand was also made. This last quantity is certainly a rather subjective parameter. The aim was to record for each stand not the tallest plant of the stand, but essentially the mean height of the vegetation in the stand as a whole. Of course, this could only be done precisely by measuring the height of every shoot in the stand – a near impossibility! So the measurement quoted for mean vegetation height for each stand was very much a subjective judgement, but is still considered to provide useful information.

LABORATORY

Back in the laboratory, the soil pH and moisture levels were determined immediately. A small sample of each soil (about 25 g) was suspended in deionized water, allowed to equilibrate for 15 minutes, and the pH determined. The remainder of each soil was weighed, dried in a paper bag to constant weight at 100 °C and reweighed: soil moisture was then calculated as

$$\text{soil moisture} = \frac{\text{fresh weight} - \text{dry weight}}{\text{fresh weight}} \times 100\%$$

The dry soil was then stored for a few weeks pending the determination of organic matter content and extractable inorganic element levels.

Organic matter content was estimated as loss-on-ignition. A sample of the soil was weighed in a nickel crucible, heated to 450 °C in a muffle furnace for 4 hours and reweighed. The calculation is:

$$\text{soil organic content} = \frac{\text{dry weight} - \text{burnt weight}}{\text{dry weight}} \times 100\%$$

For extracting inorganic elements, a solution of 1 N ammonium acetate was used, in the ratio of 1 part of soil to 5 parts of solution by weight. The mixture was shaken for 30 minutes on a mechanical shaker, filtered, and the amount of each metallic ion was determined by an atomic absorption

47

spectrophotometer. The figures obtained were parts per million of the relevant element in the solution; these were multiplied by 5 to obtain parts per million of the element in the soils, and finally division by 10 converted these latter figures into milligrams of element per 100 g of dry soil – the conventional modern unit in soil analysis.

Although it is normally recommended that soil phosphorus be extracted in 1 N acetic acid (see p. 61), the ammonium acetate solutions were also used for this element for expediency. The method employed was based on the quantitative formation of the blue compound ammonium phospho-molybdate, and the amount of this compound formed in each case was estimated by an absorption spectrophotometer. Full details of methods for soil analysis are given in Allen *et al.* (1974).

General description of the environment and vegetation along the transect

A major advantage of a transect, as opposed to a two-dimensional grid, method of sampling is that it is easy to describe environmental and vegetational changes along it.

ENVIRONMENT

The highest point, at 0 m, is 6.4 m above the lowest point which is situated at 100 m along the transect. From a uniform relatively steep slope down to 50 m there is a sudden transition to a gentler slope from 50 to 90 m along the line. Over the remaining 30 m, there is level ground for the first 10 m, followed by a slight rise over the remaining 20 m (Fig. 4.1a).

The soil moisture and organic contents are positively correlated on the large scale, and together are negatively correlated with height of ground. The two extremes of soil moisture and organic matter are over the first 20 m on the highest ground (around 10%), and between 90 and 110 m in the basin at the bottom of the transect (around 80%) (Fig. 4.1b). Despite the irregularities, there is a clear fall in soil pH of about half a unit down the transect (Fig. 4.1c).

The extractable soil inorganic element levels are shown in Figures 4.2 and 4.3, and may be considered together. Levels of all nutrients are uniformly very low over the first 20 m of the transect and, in general, are high in the wet area between 90 and 110 m with a slow rise between 20 and 90 m; but there are many variations on this general trend. For example, calcium and magnesium (Figs 4.2a,b) are conspicuously high between 30 and 50 m, and from 70 to 75 m; while calcium shows two peaks between 90 and 100 m, and magnesium has a general high level between 90 and 110 m. The magnesium: calcium ratio can be an environmental feature of ecological importance, particularly where it achieves very high values in soils over

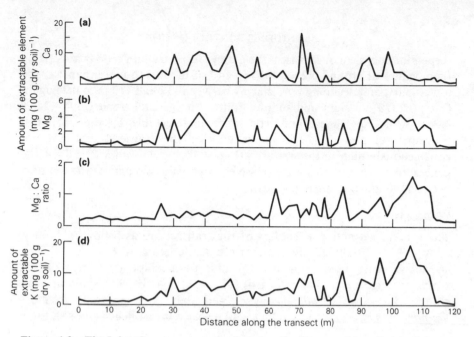

Figure 4.2 The Iping Common transect line. Levels of extractable soil: (a) calcium; (b) magnesium; (d) potassium. In (c) is shown the magnesium : calcium ratio.

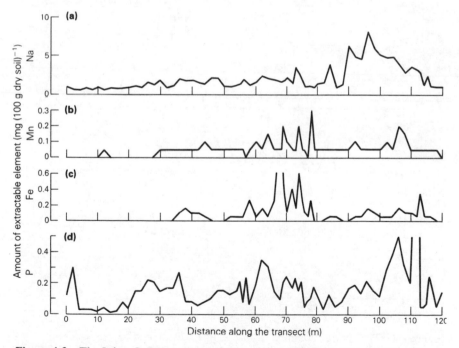

Figure 4.3 The Iping Common transect line. Levels of extractable soil: (a) sodium; (b) manganese; (c) iron; (d) phosphorus.

serpentine rocks. In our transect the magnesium : calcium ratio is low, rising from values of around 0.25 at the top end of the line to more than 1 at the lowest part, and declining very sharply between 112 and 113 m to 0.25 again (Fig. 4.2c). Potassium and sodium follow the general trend (Figs 4.2d & 4.3a), while manganese and iron were undetectable by the analytical procedure employed at the upper end of the transect, and are not conspicuously high in the wet area (Figs 4.3b, c). Phosphorus follows the general trend, but attains some relatively high values in part of the wet area between 90 and 110 m (Fig. 4.3d).

VEGETATION

The general vegetational features of the transect are as follows. The first 32 m are essentially through 8-year-old Callunetum with various lichen species growing on the soil over the first 14 m (epiphytic lichens on the *Calluna vulgaris* were not recorded) (Figs 4.4a & 4.5d). The stand at 12 m contained very little *Calluna vulgaris* and about 40% cover of the moss *Campylopus interoflexus*, together with a scattering of lichens; hence, there

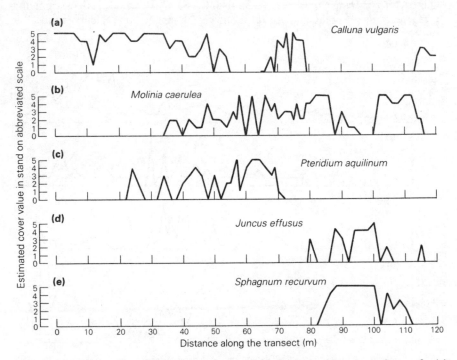

Figure 4.4 The Iping Common transect line. Estimated cover values of: (a) *Calluna vulgaris*; (b) *Molinia caerulea*; (c) *Pteridium aquilinum*; (d) *Juncus effusus*; (e) *Sphagnum recurvum*. The scale of cover values is that given in Table 1.4.

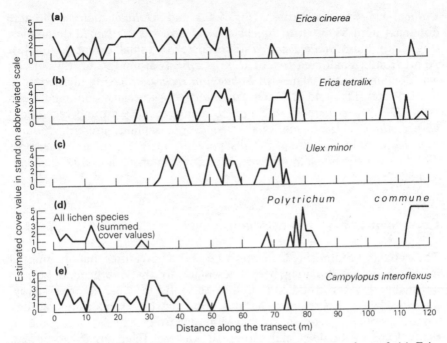

Figure 4.5 The Iping Common transect line. Estimated cover values of: (a) *Erica cinerea*; (b) *Erica tetralix*; (c) *Ulex minor*; (d) all lichen species in the data set and *Polytrichum commune*; (e) *Campylopus interoflexus*. The scale of cover values is that shown in Table 1.4.

was much bare ground here (Figs 4.1a, 4.4a, 4.5d,e). From 14 to 20 m the *Calluna vulgaris* showed an increased vigour, but not a concomitant increase in cover; however, the only environmental factor correlating with this was an increased soil pH (Fig. 4.1c). At 18 m there was a wood ants' nest within the stand, which necessitated wariness when sampling! This stand showed the highest soil pH on the whole transect (Fig. 4.1c). Between 22 and 30 m there was a narrow band of birch trees (*Betula* spp.) and some bracken (Fig. 4.4c).

From 36 to 48 m there was a zone dominated by birch and bracken, with most of the stands also containing *Ulex minor* (dwarf furze) (Fig. 4.5c). The grass *Molinia caerulea* (purple moor-grass) now entered the vegetation, and was almost continuously present along the remainder of the transect (Fig. 4.4b). The bracken was sparse and low between 50 and 56 m, but from 58 to 70 m it was tall and vigorous (Fig. 4.4c). In general, this was also a birch zone, but few appeared in the transect stands.

A series of changes occurred from 70 to 86 m. Up to 78 m, *Calluna vulgaris* and *Molinia caerulea* co-dominated, but from this point onward

Calluna vulgaris disappeared (Fig. 4.4a) and *Molinia caerulea* became dominant until 88 m when *Juncus effusus* (soft rush) attained dominance (Fig. 4.4d); but *Molinia caerulea* still occurred to a small extent (Fig. 4.4b). Apart from a small area free of *Juncus effusus* around 92 m, this species was dominant until 100 m with *Sphagnum recurvum* having high cover at ground level (Figs 4.4d,e). Then *Molinia caerulea* became dominant again until 112 m when there was a complete change to a dry ground vegetation: with *Calluna vulgaris* and small *Betula* spp. saplings present, *Molinia caerulea* rapidly decreasing and disappearing after 115 m, and the moss *Polytrichum commune* having very high cover at ground level (Figs 4.4a,b & 4.5d).

Coed Nant Lolwyn – a deciduous wood

The woodland comprising Case Study no. 2 shows features that are unusual in mid-Wales. It has long been known locally that the ground flora is unusually rich for woodland in the Aberystwyth area. A tree survey, described below, has revealed that this wood may be designated as ash–oak wood, although it is obviously very different from the ash–oak woods of the calcareous boulder clay areas of eastern England. Nevertheless, although the wood is small, the detailed study has revealed a wide range of levels of several environmental factors; hence this wood is ideal for studying the relationship between species distributions and environmental factors, and this was the subject of a research project in 1976 by Dr Elizabeth A. Wolfenden (Wolfenden 1979). In this chapter, the vegetation and its environment will be introduced, and the results of a tree survey described. In Chapter 9, ordination results will be used to assess the overall structure of the vegetation in the study area, and also to erect hypotheses about the distributions of individual species in relation to environmental factors. A formulation of the species assemblages will also be attempted using classification methods; and, in keeping with the main aim of the project, the associations produced by the classifications will be examined from an environmental, as well as a vegetational, viewpoint.

The study area

Coed Nant Lolwyn (Llolwyn Brook Wood) is situated about 5 km south of Aberystwyth and 2.5 km inland from Cardigan Bay (National Grid Reference SN5876). Like most deciduous woodlands in mid-Wales, Coed Nant Lolwyn is situated on the sides of a river valley, and at its broadest the wood is only about 0.1 km wide. The river and its valley follow a sinuous course, and a length of about 0.5 km on one side of the river – the southern

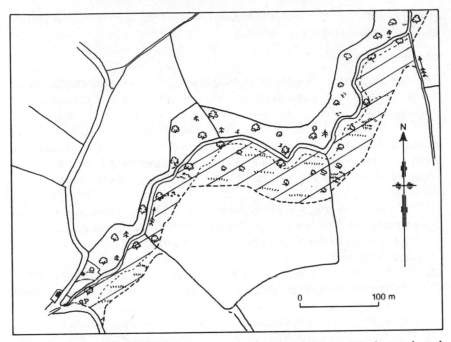

Figure 4.6 Map of part of Coed Nant Lolwyn. The study area, to the south and east of the river, is shown hatched. The map is based on the 1905 edition of the 1 : 2500 series.

side – comprised the study area (Fig. 4.6). It is this part of Coed Nant Lolwyn that contains the varied species assemblages already referred to; the remainder of the wood contains less of ecological interest.

Rocks of the Silurian era outcrop over a large part of mid-Wales, including the Aberystwyth area. Soils of the Denbigh Series (mull phase) occur on either side of the Llolwyn Valley, and the soils of Coed Nant Lolwyn itself are undifferentiated valley soils derived from material on the surrounding higher ground (Rudeforth 1970).

A study of the species list of the ground flora of the study area (see Table 4.5) suggests that the woodland is old and well established. Species of poor colonizing ability, regarded by Peterken (1974) as indicators of primary woodland in central Lincolnshire, are present in quantity in Coed Nant Lolwyn: i.e. *Milium effusum* (wood millet), *Oxalis acetosella*, *Chrysosplenium oppositifolium* (golden saxifrage), *Melica uniflora* (wood melick), *Anemone nemorosa*, *Lysimachia nemorum* (yellow pimpernel), *Adoxa moschatellina* (moschatel), *Sanicula europaea* (wood sanicle), *Veronica montana* (wood speedwell), *Conopodium majus* (pignut), and *Primula vulgaris* (primrose). *Mercurialis perennis*, recognized by Rackham

(1971) as indicative of ancient woodland in Norfolk, is a prominent feature of Coed Nant Lolwyn's ground flora.

HISTORICAL NOTES

The sole reliance on species as indicators of the historical continuity of woodland has been questioned by Streeter (1974). He maintained that documentary evidence is necessary to confirm the history of a site. The southern and eastern side of Coed Nant Lolwyn was part of the Nanteos estate, and so the history can be traced from detailed estate maps, starting from the eighteenth century. The following datelines have been determined from records held in the National Library of Wales, Aberystwyth:

(1) In 1764, Brynecrwyn Farm in the parish of Llanychayrne was surveyed and mapped. The evidence points to the existence of woodland following the valley of the Nant Lolwyn, but no definite woodland boundaries were delineated.

(2) The first large-scale map of the Aberystwyth district of 1803 (on which subsequent maps, 1820 and 1833, were based) shows the continued existence of Coed Nant Lolwyn. The small size of the woodland and the map scale do not lend themselves to accurate boundary details.

(3) The Nanteos estate was surveyed and mapped in 1819. Again, boundaries between woodland and farmland are not marked; but woods are mapped by tree symbols, and it is interesting to note that the southwestern end of the wood is denoted as sparsely wooded.

(4) The tithe map of Llanychaiarn parish in 1845 gives the total area of woodland as 8.5 acres. All previous maps show Coed Nant Lolwyn as completely wooded up to the southwestern boundary (but see 3 above), but the 1845 tithe map shows that no woodland existed between the small road and the first parallel field boundary in the southwestern corner (Fig. 4.6). No definite conclusion can be drawn from this; it may mean that the area was once wooded and felled before 1845, or that inaccuracies of earlier maps failed to illustrate the true extent of woodland.

(5) The second edition 1:2500 Ordnance Survey series shows the extent of woodland in 1905, and the relevant part of the map is reproduced in Figure 4.6. Woodland extended right to the southwestern boundary as it does today, but only in a very narrow zone by the river. A somewhat broader area between the road and first field boundary appears to be designated as 'rough pasture'.

Overall, then, Coed Nant Lolwyn can be traced back in an unbroken sequence to at least 1764, but the exact extent of the woodland boundaries has probably changed somewhat. In 1905 (Fig. 4.6), not only did the

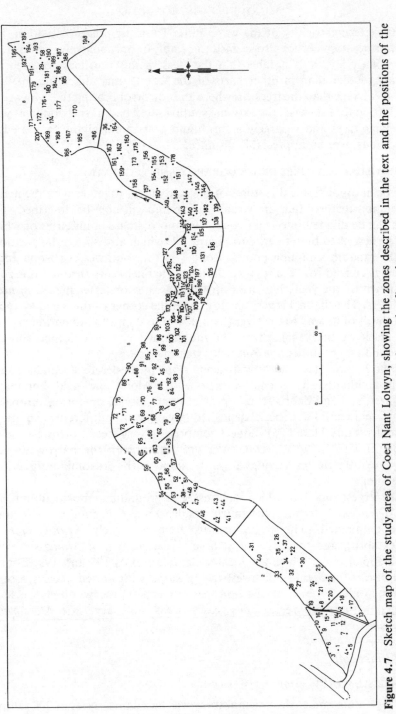

Figure 4.7 Sketch map of the study area of Coel Nant Lolwyn, showing the zones described in the text and the positions of the random sampling points. The zone numbers are shown above (to the north of) each zone.

southwestern extremity of the wood differ from the present situation, but the northeastern corner also seemed to be not, or only sparsely, wooded. In this connection, it is notable that the trees in this northeastern area are mostly smaller than in other parts of the wood, thus indicating they are younger in age than the trees elsewhere and so corroborating the evidence of the 1905 map. Likewise, the extreme southwestern end of the wood has very few large trees and is largely a shrub and grass area, giving 'wood edge' conditions over an appreciable distance.

MAIN VISUAL FEATURES OF TOPOGRAPHY AND VEGETATION

The main topographical features of the study area, together with the major vegetation features that are visually apparent, will now be described. The area can be divided into eight zones, the approximate boundaries of which can be described by reference to Figure 4.7, which also shows the positions of the random sampling points. Zone 1 is the southwestern wood edge, already referred to. A large field drainage channel flows through this zone for much of the year, but dries up for a greater or lesser period in most summers. The channel is very shallow from the corner of the wood to about two-thirds of its way to the river (Sample Points 2 and 14 are on wet ground for much of the year), thereafter the channel is much deeper and the surrounding ground is not especially wet.

Zone 2 comprises a northwest-facing slope of moderate steepness, much of it dominated by *Mercurialis perennis*. Along the field boundary, however, soil and debris from a previous agricultural operation (probably hedge removal) have been deposited giving quite different conditions (Sample Points 23 and 25). Zone 3 comprises a very steep northwest-facing slope, dominated by *Hyacinthoides non-scripta*, but the narrow Zone 4 is quite different in having a gentle slope with predominantly grassy vegetation.

The largest zone is 5, which is almost level ground and mostly dominated by *Mercurialis perennis*; the much smaller Zone 6 is similar but mainly shrub-dominated. The scrub in this zone is largely *Prunus spinosa* (blackthorn), and there is a very large spreading tree of *Malus sylvestris* (crab apple) under which are located Sample Points 197 and 199. Zone 7 comprises a moderate northwest-facing slope with varied species assemblages; while Zone 8, at the northeastern end of the wood, is on level ground with small trees, some grassy areas and very little *Mercurialis perennis*.

Field methods

ESTABLISHING THE SAMPLE POSITIONS

The vegetation and environmental study was based on a series of 200 sample points within the area delineated in Figure 4.6. The positions of the points were randomly determined, by the procedure to be described below, since the main objective of the work was to erect hypotheses concerning the effects of the measured environmental factors on species distributions by various statistical analyses.

The boundaries of the wood were plotted using enlargements of aerial photographs, the scale of which was checked against the 1 : 2500 map series (1905, 2nd edn). Within this framework, the positions of selected trees, in relation to each other and to the woodland boundaries, were fixed by compass triangulation. Difficult terrain within the wood precluded the use of other, more accurate, survey methods. These marker tree positions were plotted onto an outline map of the wood, a grid system superimposed, and 200 sample points were finally positioned on the map using random number tables (Fig. 4.7). The distance and bearing from the nearest marker tree to each sample point were measured on the map, and these measurements, translated to the wood, enabled each point to be located on the ground and permanently marked by a numbered wooden stake driven into the soil. From Figure 4.7, it can be seen that, in general, the sample point numbers increase from south-west to north-east of the study area. The few discrepancies in this trend were due to sample points that were ultimately found to be located outside the wood, owing to small errors in map drawing! These were re-randomized, and could thus appear anywhere within the area.

TREE SURVEY

A tree survey was carried out by the point-centred quarter method (p. 29), based on a subset of the 200 random sampling points. Not all the points were used, since some were so close together that substantially the same trees would have been selected from each of a number of points. At each location, the area around was divided into quarters, based on the main points of the compass, and within each quadrant the nearest tree (greater than about 6 m high) was identified, its distance from the point measured, and the circumference of the tree at breast height recorded. ('Circumference of the tree at breast height' is a term used by foresters, being 1.5 m above ground level.)

One hundred and fifty sample points were actually used, involving 540 tree measurements (the full quota of 600 possible tree measurements was not obtained, because in some quadrants there was no tree between the sample point and the woodland boundary). From the measurements, the following quantities were calculated from the formulae of Curtis & McIntosh (1950):

$$\text{relative density of species X} = \frac{\text{density of species X}}{\text{total density of all species}} \times 100$$

57

relative frequency of species X

$$= \frac{\text{frequency of species X}}{\text{sum of frequency values of all species}} \times 100$$

relative dominance of species X

$$= \frac{\text{sum of circumferences for species X}}{\text{sum of circumferences for all species}} \times 100$$

Dealing with the circumference measurements was straightforward, while the frequency of species X is given by the actual number of sample points which records at least one individual of that species. The calculation of the density of species X is more complicated. Let l be the mean of all the 540 distance measurements made; then density, d, of all trees in the wood is estimated by $d = 1/(l)^2$. The density of species X is then calculated by multiplying d by the total number of records obtained for species X divided by the total number of tree records (540). Finally, the importance value for species X is calculated by summing its relative density, relative frequency and relative dominance.

GROUND VEGETATION SURVEY

The 200 random sample points were used for defining the positions of stands in the ground vegetation survey. A square quadrat of area 0.25 m^2 was used. A smaller stand would have introduced excessive spatial exclusion and edge effect in relation to the size of many woodland angiosperm species. A 1 m^2 square quadrat was also tried, but proved impossible to manipulate under conditions of thick undergrowth and steep terrain. Also, as the main aim of this work was to correlate species occurrences with particular environmental factors, the smaller stand size was preferred as environmental parameters can change markedly over short distances.

Qualitative data were recorded for all species, and quantitative (density) measurements were made for all ferns and angiosperms except *Hedera helix* (ivy) and *Lonicera periclymenum* (honeysuckle). In many respects, density is the best abundance measure, but it has the disadvantage that the definition of an 'individual' must depend on the morphology of the species concerned. In this study, an 'individual' ranges from a complete plant down to a single leaf growing directly up from a rhizome; details are given in Table 4.5. Because of the phenological succession occurring on the woodland floor (Al-Mufti *et al.* 1977), the timing of density observations is critical. The time of year when each species is at maximum density differs. To overcome this difficulty, a sequence of recording times was instituted between the beginning of March and the end of September. It was thus hoped that the maximum density for the year's growth was recorded for each species in every stand in which it occurred.

ENVIRONMENTAL FACTORS

Light The light meters employed were designed in the Department of Botany & Microbiology at Aberystwyth. They are 'integrating' light meters; so that, rather than giving an instantaneous light value at any one time, they integrate the total light received over a specified period of time. The light meters are photochemical in nature: the light, suitably filtered to screen out ultraviolet and much of the blue light to which the light-sensitive fluid (photolyte) responds, is allowed to enter an opaque jar containing the photolyte (potassium ferrioxalate) which then undergoes a chemical change, the amount of which is proportional to the total amount of light absorbed. Full details may be found in Wolfenden *et al.* (1982).

A large quantity of the photolyte was made up at least two days before the proposed time of exposure, and stored in the dark in an amber bottle. The day before exposure in the wood, 25 cm^3 of photolyte was put into the jar of each meter, then stored and transported to the site in darkness. At every sample point, a light meter was placed above the field layer but below the shrub layer. Although it took a finite time interval to distribute the meters and to gather them up two days later, this was always done between 1100 and 1300 hours; and so the exposure time was about 48 hours.

It was impossible to find a safe and completely unshaded site outside the wood to expose a batch of meters for the full daylight reading. A group of 12 meters was therefore placed on the roof of the Biology building at the University College of Wales, Aberystwyth (5 km NNE). The statistical error of the full daylight reading is thus considerably smaller than the reading for each stand, but there may be an overall bias of the full daylight relative to the woodland results owing to their distance apart. The light result for each stand was expressed as a percentage of full daylight. The details of the four exposure periods are:

(1) first reading (pre-vernal) – before the tree canopy opened, 6–8 April 1976;
(2) second reading (vernal) – canopy expansion begun, 10–12 May 1976;
(3) third reading (aestival) – full expansion of canopy, leaves young, 7–9 June 1976;
(4) fourth reading – should have been full canopy of older leaves, but many leaves prematurely abscised because of abnormal drought conditions, 16–18 August 1976.

Soils Seasonal fluctuations in the properties of soils are frequently reported, but there is considerable disagreement on the magnitude and timing of these changes (Barker & Clapham 1939, Ball & Williams 1968, Davy & Taylor 1974, 1975). Spatial variation has also been stressed;

59

differences in chemical content over very short distances have been demonstrated by Piper & Prescott (1949), who believed that the spatial heterogeneity of the soil far outweighed any observable temporal fluctuations. This problem has been statistically examined by Frankland *et al.* (1963). On investigation of sessile oak and larch woodland soils, it was found that variations within each sample plot on different days were usually so large that it was difficult to show any temporal fluctuations. There is, however, some suggestion that inorganic phosphorus (Saunders & Metson 1971) and nitrogen (Davy & Taylor 1974) levels are higher in the spring. Overall, it would seem that temporal variation in soil nutrient levels are of lower magnitude than spatial variation.

The drought of the summer of 1976 and the resulting hardness of the ground, however, precluded soil sampling until after the onset of heavy rain in September. Moreover, the optimal time to sample soil so that moisture content at field capacity may be determined is 48 hours after heavy rain (Stewart & Adams 1968), and this was duly done. Four soil samples were taken at each of the 200 sample sites (one from each corner of the 0.25 m^2 stand), using a trowel to a depth of about 150 mm after removal of surface litter.

Laboratory methods

Light   On returning the light meters to the laboratory, the photolyte was assayed as described in Wolfenden *et al.* (1982). Solarimeter readings at the nearby Welsh Plant Breeding Station showed that exposure of the meters in full daylight for the 48-hour periods in June and August would exceed the linear response range. Accordingly, a calibration curve, Figure 4.8, was used to convert the photolyte absorbance measurements to radiation integrals (i.e. total amount of light received), and hence percentages of full daylight.

Soils   For all measurements except those of pH and moisture, the four sub-samples of soil from each sampling site were thoroughly mixed, air dried at room temperature and sieved through a mesh of 2 mm. In this state the soils could be stored until assayed.

Moisture and pH measurements, using fresh soil, were completed within a few days after field sampling. The pH of each of the four sub-samples taken from every stand was measured separately. By this means, the within-stand variation was found to be small, and so it was assumed that this variation was also small for the other soil factors measured: this was the rationale for bulking the sub-samples of soil from each site before further analysis, including the moisture assay.

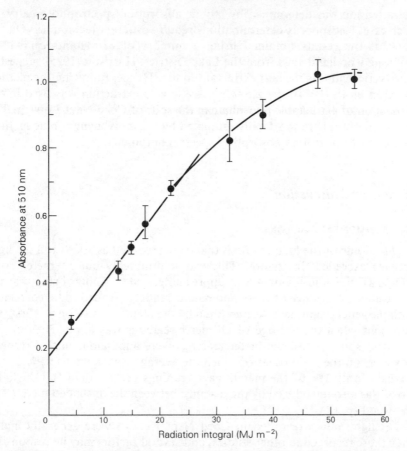

Figure 4.8 The mean absorbance value of five replicate integrating light meters at nine radiation integrals, with twice the standard error on each side of the mean (calculated for each batch separately). The linear regression line is shown for the five lower points, together with an eye-fitted curve for the upper four points (after Wolfenden *et al.* 1982, by permission of *Acta Oecologia*).

Preliminary tests These comprised pH, moisture, and organic matter assays. The methods of Allen *et al.* (1974) were followed; and so pH was measured in a sludge of the soil in deionized water, moisture by drying at 105 °C, and organic matter by loss-on-ignition at 450 °C.

Chemical analyses Ammonium acetate (pH 7) was chosen as the extractant for the determination of the cations – sodium, potassium, calcium, magnesium and manganese. The precise analytical techniques, and the reasons for their choice, are given by Allen *et al.* (1974). The amount of

61

extracted ion was determined by atomic absorption spectrophotometry, in each case. Phosphorus determinations require different techniques. On the basis of the results obtained using a number of extractants on widely differing woodland soils from the Lake District, Harrison (1975) suggested that acetic acid was the best extractant to use. Hence, following the method of Allen *et al.* (1974), the 2.5% v/v acetic acid extraction was used in the estimation of extractable phosphate in the soils of Coed Nant Lolwyn. The extracted phosphate was finally estimated by the molybdenum blue method in conjunction with an absorption spectrophotometer.

Results and discussion

ENVIRONMENTAL FACTORS

Light   Some of the light readings that were recorded as 100% full daylight actually exceeded this figure! This was undoubtedly due to meter error which, as shown in Figure 4.8, is appreciable; and it is noteworthy that the percentage occurrence of these anomalous readings tended to be correlated with the general light level as measured by the meters. Thus, on the first two occasions when the average of all meter readings was about 50% of full daylight, some 8% of the meter readings were anomalous in this respect; however, on the third occasion when the average was down to 25% of full daylight, only 1% of the meters gave readings greater than full daylight. Error was also introduced by the distance between the unshaded meters and the wood (p. 61).

The light values recorded in Coed Nant Lolwyn were generally higher than those recorded in other studies, and several factors may be responsible for this difference. The majority of investigations have used instantaneous light readings from matched pairs of photoelectric cells, one at the site within the wood and the other outside the tree canopy, and presented the site results as a percentage of the full daylight reading outside the wood. There are many disadvantages to this procedure:

(a) the individual readings of any one pair must be truly instantaneous, as light intensity is continually changing under most weather conditions even over very small time intervals;
(b) the presence of sunflecks at a point, and their duration, may affect an instantaneous light reading very markedly, but may be quite insignificant from the point of view of the general light climate of that site;
(c) the altitude of the sun in relation to the structure of the tree canopy and the slope of the ground at a woodland site will affect instantaneous light readings taken at different times of the day in a different manner from

Table 4.2 Comparison of light values from three sources. Each figure is a light value expressed as a percentage of full daylight.

Anthracene in benzene photochemical solution (Pierik 1975)

	April	April	May	May	June	June	August
alder–ash wood	47	41	31	24	13	5	1
alder wood	47	33	17	10	4	2	1
beech wood	39	34	4	1	1	1	1
oak coppice wood	55	53	36	34	8	7	3

Computed from hemispherical photographs (Anderson 1964)

	January–April		July–October	
	diffuse mean	direct mean	diffuse mean	direct mean
large clearing	33.6	12.5	25.9	2.5
small clearing	29.0	14.0	8.0	2.3

Potassium ferrioxalate photochemical solution (present investigation)

	April	May	June	August
mean of all 200 sites	50	50	25	48

the steady change in light intensity that would be expected under open sky conditions.

Clearly, an integrated light reading over at least 24 hours is required.

Table 4.2 compares the present results with those from a similar study using photochemical light meters (Pierik 1975), and with values calculated from hemispherical photographs (Anderson 1964). Apart from the initial reading in April, the results for all Pierik's sites are consistently lower than those measured in Coed Nant Lolwyn. An explanation of this difference may be gained from knowledge of the spectral sensitivities of the two photolytes involved. Potassium ferrioxalate solution will record well into the visible spectrum (Melville & Gowenlock 1964), and the design of our light meters ensured that most of the ultraviolet radiation, which would be preferentially absorbed, is filtered out. In comparison, Pierik's photolyte, anthracene, only absorbs wavelengths below 390 nm, and he further states: '... light absorption below 400 nm, by canopy leaves, is very great, almost no ultra-violet radiation will reach the undergrowth. Thus the anthracene method may be expected to be solely related to the cover density.' Similarly, the values recorded by Anderson (1964) in Madingley Wood, Cambridgeshire, are somewhat lower than the Coed Nant Lolwyn results. It is interesting to note that the values obtained for Anderson's large clearing site, for the July–October period, are closest to those of Coed Nant Lolwyn at full canopy expansion in June. This leads to the hypothesis that the high light values of the present investigation may be correlated with the configuration of the woodland site. Coed Nant Lolwyn is very narrow, and so light from the boundaries may easily infiltrate the whole wood. Further, the light readings include those from many wood edge sites, and from other areas where the trees are fairly small and with a rather open canopy.

Soils There are not many sources in the literature on the nutrient contents of woodland soils against which our soil analysis results may be compared; moreover, there is the difficulty, clearly shown by Harrison (1975) for phosphorus, that the precise extraction and measurement procedures need to be known before a rigorous comparison can be made of the results of two different investigations. However, since the publication of Allen *et al.* (1974), there has been a very marked tendency for investigators to follow the procedures contained therein: these methods are becoming standard for ecological investigations.

The seven sets of results shown alongside the means and ranges of the Coed Nant Lolwyn data in Table 4.3 may be used as comparisons, because the extraction methods were similar. The examples range from a mor soil through to various calcareous soils, and it is evident that for pH, calcium, and potassium, the soils of our study area occupy a medium position in the ranges encountered in Britain. Soils of Coed Nant Lolwyn seem to be very low in organic matter and manganese, and also in moisture (mean of 35.6%). In respect of this last factor, Harrison (1975) gave moisture values for 16 woodland soils, and these lay in the range 60–98%. In the same 16 soils phosphate, extracted with 2.5% acetic acid, gave readings between 0.31 and 0.83 mg $(100 \text{ g})^{-1}$ with only one result, 2.21 mg $(100 \text{ g})^{-1}$, higher than the Coed Nant Lolwyn soil mean for this element. The magnesium contents of the Coed Nant Lolwyn soils average out high, and sodium is very high; this could be due to the proximity of the sea, especially in relation to sodium. The loss-on-ignition seems to be extremely low; it is possible that much of the soil organic matter was oxidized during the prolonged hot and dry weather in the summer of 1976.

The picture that emerges is that the soils of the study area were relatively dry and base-rich, with above average phosphorus content. They are very much mineral rather than organic soils. The dryness, as measured by the soil moisture at field capacity, is further confirmed by the low level of extractable manganese. The manganous ion (divalent) is the most available form of this element, and this is only produced in quantity in waterlogged soils; in dry soils Mn^{2+} tends to be oxidized to unavailable (and unextractable) higher valency states (Etherington 1982).

Table 4.3 Comparison of nutrient and other soil factor levels in woodland soils from five sources. Organic matter (loss-on-ignition) is expressed as a percentage by weight of dry soil, and all elements are in mg $(100\ \mathrm{g})^{-1}$ of dry soil.

	Lake District†			Cambridgeshire‡	Chilterns§			East Scotland¶
	Coed Nant Lolwyn		Roudsea wood (mull-moder)	Hayley Wood (chalky boulder clay)	Hillock Wood (acid mull)	Combe (Calcareous brown earth)	Hobbs Hill (chalk soil)	Morrone Birkwoods (acid peat to calcareous soils)
	mean*	range		Merlewood (mor)					
pH	4.9	3.8–6.1	4.3	4.2	5.0–7.0	4.0	7.5	8.0	4.2–6.8
organic	1.4	0.3–10.2	18.4	14.0	—	—	—	—	15.6–75.5
calcium	89.4	10.9–263.8	10.0	33.0	299.6–507.2	60.0	1000	740	42.4–491.1
magnesium	20.2	4.4–45.6	4.9	5.0	19.6–24.0	3.6	9.6	4.8	12.8–39.4
potassium	16.3	1.3–59.1	5.1	5.2	23.4–42.9	15.6	7.8	5.0	6.8–91.0
sodium	35.7	1.0–229.7	2.9	3.6	4.7–6.5	—	—	—	6.5–10.3
manganese	1.2	0.1–5.4	—	—	19.9–24.3	8.3	undetectable	—	—
phosphorus	1.0	0.5–4.3	0.5	0.8	—	—	—	—	—

* Geometric means for elements, arithmetic means for pH and organic matter.

† After Frankland et al. (1963). Figures are for early October.

‡ After Martin & Pigott (1975). Figures for pH, Mg^{2+}, K^+, Mn^{2+} are for the 0–5 cm depths. Those for Ca^{2+} and Na^+ are means for depths over the range 0–20 cm, since depth readings for these elements are not significantly different from one another.

§ After Davy & Taylor (1975).

¶ Huntley & Birks (1979). The figures reproduced here are calculated from the following formulae:

$$\text{Weight of element, mg }(100\ \mathrm{g})^{-1}\text{ soil} = \frac{(\text{weight of element, mEq l}^{-1})(\text{equivalent weight of element})}{(10)(\text{soil bulk density, g cm}^{-3})}$$

where soil bulk density, g cm^{-3} = $1.482 - 0.6786 \cdot \log$ (percentage loss-on-ignition) (Jeffrey 1970).

VEGETATION

Trees   The tree survey results are given in Table 4.4. *Fraxinus excelsior* (ash) has the highest importance value, and so is the leading dominant in this stand of trees. Two other species have reasonably high importance values, namely *Quercus petraea* (sessile oak) and *Acer pseudoplatanus* (sycamore). It is due to *Quercus petraea* having the second highest importance value that justifies classifying the area as an ash–oak wood, while *Acer pseudoplatanus* is a very invasive species in mid-Wales woods. Interestingly, *Crataegus monogyna* (hawthorn) appears next in the ranking of importance values, although with a figure of much lower magnitude. This is probably a reflection of the age of the wood, in that from the shrub layer and incorporated scrubland many individuals of *Crataegus monogyna* have had sufficient time to grow and become a notable feature of the tree layer.

Fraxinus excelsior occurs where the soil is particularly moist or in areas where the soil is relatively base rich. The composition of the ground flora, particularly the abundance of *Mercurialis perennis*, indicates that at least the latter condition holds for large areas in this wood. The dominance of ash, and the relative abundance of *Mercurialis perennis* provide evidence, along with that presented in the discussion on soils above, that the study area has a relatively base-rich and mature woodland soil.

Table 4.4   Results of the tree survey in Coed Nant Lolwyn. The overall density of trees is 0.029 m^{-2}.

	Relative density	Relative frequency	Relative dominance	Importance value
Fraxinus excelsior	34.07	29.17	34.42	97.66
Quercus petraea	22.96	20.83	27.20	70.99
Acer pseudoplatanus	18.70	21.11	19.06	58.87
Crataegus monogyna	8.33	8.89	6.64	23.86
Ulmus glabra	6.30	8.06	5.26	19.62
Prunus avium	3.33	3.06	2.62	9.01
Fagus sylvatica	1.48	2.22	1.44	5.14
Corylus avellana	1.67	2.22	1.03	4.92
Betula pendula	1.48	1.94	1.05	4.47
Salix spp.	1.11	1.67	0.78	3.56
Malus sylvestris	0.56	0.83	0.50	1.89

Ground vegetation   One hundred species of flowering plants (including the seedlings of woody species), ferns and bryophytes were recorded in the 200 stands. The species are listed, together with their relative frequencies of occurrence, in Table 4.5. Density values, where recorded, were quoted for

Table 4.5 The species of the field and ground layers (including tree and shrub seedlings) recorded in the 200 sample stands in Coed Nant Lolwyn. The abbreviations adopted in text figures and tables elsewhere in the book, the percentage frequency of occurrence in the stands and maximum density (where applicable) are also given for each species.

	Abbreviation	Percentage frequency	Maximum density (m^{-2})
(a) species in which the individual plant is recognizable and used for density measurements			
Adoxa moschatellina	Am	11.5	260
Alliaria petiolata		1.0	32
Cardamine pratensis		1.0	16
Chaerophyllum temulentum		1.0	16
Circaea lutetiana	Cl	35.0	84
Conopodium majus	Cm	6.0	40
Epilobium montanum	Em	3.5	68
Filipendula ulmaria	Fu	8.0	96
Fragaria vesca	Fv	3.0	16
Geranium robertianum	Gr	22.5	144
Geum urbanum	Gu	35.0	72
Heracleum sphondylium	Hs	7.0	8
Hyacinthoides non-scripta	Hn	42.5	224
Hypericum pulchrum		0.5	4
Lapsana communis	Lc	2.0	8
Lathyrus montanus		1.0	20
Listera ovata		0.5	12
Lysimachia nemorum		0.5	96
Moehringia trinervia	Mt	4.0	36
Plantago lanceolata		0.5	4
Potentilla sterilis	Ps	8.0	32
Primula vulgaris	Pv	7.0	12
Ranunculus ficaria	Rf	70.0	224
Rumex acetosella		0.5	4
Rumex crispus	Rx	1.5	4
Sanicula europaea	Se	16.5	28
Solidago virgaurea	Sv	2.0	48
Stachys betonica	Sb	2.0	16
Taraxacum officinale	To	6.5	8
Trifolium pratense		0.5	20
Viola riviniana	Vr	10.5	56
Acer pseudoplatanus	Ap	8.0	16
Crataegus monogyna	Cg	3.5	4

Table 4.5 (*continued*)

	Abbreviation	Percentage frequency	Maximum density (m^{-2})
Fraxinus excelsior	Fe	12.5	24
Humulus lupulus		0.5	8
Prunus avium		0.5	4
Prunus spinosa	Pp	4.0	12
Athyrium filix-femina	Af	1.5	4
Dryopteris affinis		1.0	4
Dryopteris dilatata	Dd	8.5	8
Dryopteris filix-mas	Df	3.5	4
Phyllitis scolopendrium		1.0	4
(b) species in which the individual plant is recognizable but density is the number of aerial stems			
Galium aparine	Ga	63.5	320
Prunella vulgaris		1.0	8
Veronica chamaedrys	Vc	14.0	204
Veronica hederifolia	Vh	5.0	200
Veronica montana	Vm	31.5	84
(c) species in which the individual plant is not, or not readily, recognizable and density is the number of aerial stems			
Ajuga reptans		1.0	16
Glechoma hederacea	Gh	5.0	104
Mercurialis perennis	Mp	40.5	268
Silene dioica	Sd	16.0	244
Stachys sylvatica	Ss	2.5	32
Stellaria holostea	Sh	21.0	376
Teucrium scorodonia		0.5	4
Urtica dioica	Ud	16.0	160
Rosa canina	Rc	4.0	24
Rubus fruticosus	Ru	15.0	16
Rubus idaeus	Ri	2.0	8
Pteridium aquilinium		0.5	4
(d) species in which density is the number of flowering stems[*]			
Agrostis capillaris	Ac	1.5	52
Anthoxanthum odoratum	Ao	2.0	100
Arrhenatherum elatius	Ae	2.5	24

Table 4.5 (*continued*)

	Abbreviation	Percentage frequency	Maximum density (m^{-2})
Brachypodium sylvaticum	Bs	7.5	76
Dactylis glomerata	Dg	5.0	44
Festuca gigantea	Fg	1.5	8
Festuca rubra		1.0	8
Holcus mollis	Hm	16.0	24
Luzula campestris	Lz	2.0	40
Melica uniflora	Mu	12.5	28
Milium effusum	Me	7.5	52
Poa nemoralis	Pn	6.0	16
Poa pratensis		0.5	4
Poa trivialis	Pt	7.5	32
Chrysosplenium oppositifolium	Co	6.5	800

(e) species in which density is the number of flowering stems and individual leaves borne directly on the rhizome

Anemone nemorosa	An	79.0	1148
Oxalis acetosella	Oa	34.5	656

(f) species in which no density measurements were made

Hedera helix	Hh	27.0	
Lonicera periclymenum	Lp	3.5	
Atrichum undulatum	Au	5.5	
Brachythecium rutabulum	Br	7.0	
Brachythecium velutinum	Bv	4.0	
Calliergon cuspidatum		0.5	
Cirriphyllum piliferum	Cp	3.0	
Eurhynchium praelongum	Ep	65.5	
Eurhynchium striatum	Es	46.0	
Fissidens bryoides	Fb	2.0	
Fissidens taxifolius	Ft	2.0	
Hookeria lucens		1.0	
Hypnum cupressiforme		0.5	
Lophocolea bidentata	Lb	6.5	
Mnium hornum	Mh	10.0	
Plagiochila asplenioides	Pa	15.5	
Plagiomnium undulatum	Pu	25.5	
Plagiothecium denticulatum	Pd	3.0	
Plagiothecium nemorale	Pl	7.5	

Table 4.5 (*continued*)

	Abbreviation	Percentage frequency	Maximum density (m^{-2})
Polytrichum formosum	Pf	2.0	
Rhynchostegium confertum	Rt	7.0	
Rhytidiadelphus loreus	Rl	2.0	
Thamnobryum alopecurum	Ta	2.5	
Thuidium tamariscinum	Tt	17.5	

*For the grass species, a density of 4 denotes the species as present but not flowering, 8 denotes one flowering stem, 12 denotes two flowering stems, etc. in a 0.25 m² quadrat.

each species in each stand as the maximum found during the several recording occasions in 1976 (p. 58). The maximum density recorded for each species is also quoted in Table 4.5. It was considered impractical to attempt recording bryophyte densities. Neither were densities recorded for the two 'trailing' species *Hedera helix* and *Lonicera periclymenum*, because frequently they occurred in a stand and were recorded as such, but may or may not have been rooted within it; thus, the interpretations of density values would not be straightforward for these species.

In contrast to the majority of woodlands in western Britain, the study area of Coed Nant Lolwyn is not bryophyte-rich, neither in terms of species nor their abundance (subjectively noted during the ground flora survey work). This may be simply due to the luxuriance of the angiosperm vegetation with consequent suppression of the bryophytes. Another unique feature of the ground flora, and again an indicator of base-rich soils, is the near absence of the calcifuges *Deschampsia flexuosa* and *Digitalis purpurea*. The former probably occurs in nearly every wood in the locality, and is abundant to dominant in many of them, while the latter is a common hedgerow species in the vicinity and could well have occurred over the extensive edges of Coed Nant Lolwyn.

Sufficient has now been written to introduce the aims of the work and the characteristics of the study area, particularly its levels of soil environmental factors in relation to other published results of British woodlands. The main purpose of the work – analyzing the ground flora – will be pursued in Chapter 9.

5

Association between species and similarity between stands

Any classificatory or ordination method comprises two components: a measure of association between species or similarity (or dissimiliarity) between stands, and a strategy of manipulation of the chosen measurement. Moreover, many of the methods required in more detailed ecological work, not involving classification or ordination, use the same measures of association or similarity, particularly those which are amenable to statistical tests of significance. This chapter is concerned with the formulation and use of such measures, while the next two chapters will deal, respectively, with classification and ordination methods.

Concepts

Association between species

Species association is, broadly, the concept of what company a species keeps, and to what degree, with other species. As an example, let us consider heathland. Of the three common heathers in Britain, *Erica cinerea* (bell-heather) tends to occur in the driest parts of heath, *Erica tetralix* (cross-leaved heath) and the grass *Molinia caerulea* in damp areas, and Bog Mosses (*Sphagnum* spp.) in the wettest parts. If one examines a small area of damp heath, there would be a high probability of both *Erica tetralix* and *Molinia caerulea* occurring together: we say that they are positively associated. On the other hand, we would almost never find *Erica cinerea* in damp heath, and only rarely would we find *Erica tetralix* or *Molinia caerulea* in the driest areas where *Erica cinerea* grows: thus *Erica cinerea* is said to be negatively associated with *Erica tetralix* and *Molinia caerulea*. A third situation is illustrated by the distribution of *Calluna vulgaris*. Although showing a preference for dry heath, this species is very widespread through many kinds of heathland soils, being absent only from the

71

wettest. Thus, *Calluna vulgaris* would usually show a negative association with *Sphagnum* spp., but not show a marked positive or negative association with the other species mentioned; that is, no association.

The above discussion has been written in the context of qualitative data, so even if just one plant of *Erica tetralix*, consisting of a single stem, occurred in a sample stand overwhelmingly dominated by *Calluna vulgaris*, we should still say that the two species occurred with equal weighting in that stand. Similar ideas extend to quantitative data; but here the problem of an excessive number of zeros, discussed in Chapter 3, can raise complications, and should be borne in mind.

Similarity between stands

Quite simply, the similarity between stands refers to their floristic composition. For qualitative data, a similarity measure between two stands gives information about the number of species common to both, relative to the total number of species occurring in one stand or the other. The more species in common, the higher the degree of similarity, so that two stands containing exactly the same species are as similar as they can be. With quantitative data there is the extra factor of the abundance of each species to consider; two stands are not identical just because they contain the same species, each species would also need to have the same abundance in each stand.

Qualitative data

Chi-square as a measure and test statistic of association between species

To measure and test for association between species on a presence or absence basis, the statistical method of contingency tables, employing χ^2, is very commonly used. To describe the method in general terms, consider the situation where two species, A and B, occur in some of a total number, n, of sample stands. The data can be arranged in the form of a 2×2 table, as shown in Table 5.1, with ' + ' representing species presence and ' − ' representing species absence. The upper left-hand cell of the four in the main body of the table shows the number of stands, a, out of the total of n in which both Species A and B are found. The lower left-hand cell gives the number of stands, b, in which Species A but not B occurs. The upper right-hand corner gives c stands containing Species B but not A, and the lower right-hand corner shows that d stands out of the n contain neither Species A nor B. Of the mariginal totals, $a + b$ gives the total number of stands containing Species A, and $a + c$ the total number of occurrences of

Table 5.1 A 2×2 contingency table for assessing association between two species, A and B.

		Species A		
		+	−	
Species B	+	a $E(a)$	c $E(c)$	$a + c$
	−	b $E(b)$	d $E(d)$	$b + d$
		$a + b$	$c + d$	$n = a + b + c + d$

Species B; conversely, $c + d$ shows the number of stands not containing Species A, and $b + d$ the total number of absences, out of n, of Species B. The extreme lower right-hand side of the table, obtained by summing either the column marginal totals at the bottom of the table or the row marginal totals on the right-hand side, is equal to the total number of stands in the data set, $n = a + b + c + d$.

Now, if there was genuinely no association between the two species, i.e. if the occurrence of one species neither increased nor decreased the probability of the other species occurring, then the probability of both species being present in a stand would be equal to the product of their individual probabilities of occurrence (the multiplication law of probability for independent events). Probabilities, however, relate to infinite populations rather than finite numbers of samples; but in the latter, relative frequencies have similar properties to probabilities. Now, the relative frequency of occurrence of Species A is $(a + b)/n$, so we can say that our estimate of the probability of the occurrence of Species A in a single stand is given by:

$$p(A) = \frac{a + b}{n} \tag{5.1}$$

Likewise, for Species B:

$$p(B) = \frac{a + c}{n} \tag{5.2}$$

and so the probability of finding both Species A and B in one sample stand is given by the product of Equations 5.1 and 5.2:

$$p(A \ \& \ B) = \frac{(a + b)(a + c)}{n^2} \tag{5.3}$$

73

However, we are interested in the *number* of stands containing both species under the assumption of no association between them, rather than the probability of finding both species in one stand under the same assumption. This number can be found simply by multiplying the probability of Equation 5.3 by n, the total number of stands involved; and so, we have

$$E(a) = \frac{(a+b)(a+c)}{n} \tag{5.4}$$

where $E(a)$ is read as the 'expected value of a', that is, the expected number of stands containing both species under the no association assumption.

It will be noticed that the expected value of a is calculated by multiplying together the two marginal totals involved (i.e. the row marginal total containing a and the column marginal total likewise) and dividing by n. A similar calculation could be made for the other three cells in the table, e.g. $E(b) = (a+b)(b+d)/n$, but with actual numerical data there is a simpler way of obtaining the other expected values once $E(a)$ has been found. This is because $E(a) + E(b) = a + b$, $E(a) + E(c) = a + c$, $E(b) + E(d) = b + d$, and $E(c) + E(d) = c + d$; that is, not only must the actual observations in a row or column of the table sum to give the appropriate marginal total, but the expected values must do so as well. Thus, we have

$$E(b) = a + b - E(a)$$

$$E(c) = a + c - E(a)$$

$$E(d) = b + d - E(b)$$

Now, for a general contingency table, χ^2 is given by

$$\chi^2_{[(r-1)(c-1)]} = \sum_{i=1}^{r} \sum_{j=1}^{c} \frac{\{x_{ij} - E(x_{ij})\}^2}{E(x_{ij})} \tag{5.5}$$

where r is the number of rows in the table, c is the number of columns, x_{ij} is the observed value in the ith row and jth column, $E(x_{ij})$ is the corresponding expected value, and $[(r-1)(c-1)]$ is the degrees of freedom. The double summation sign is interpreted thus: i is set equal to 1 and j is allowed to take successive values from 1 to c, then i is set to 2 and j again takes successive values from 1 to c. Each combination of values of i and j gives a term of the form

$$\{x_{ij} - E(x_{ij})\}^2 / E(x_{ij}).$$

For the 2×2 contingency table, where $r = c = 2$, and with the notation of

Table 5.1, Equation 5.5 reduces to

$$\chi^2_{[1]} = \frac{\{a - E(a)\}^2}{E(a)} + \frac{\{b - E(b)\}^2}{E(b)} + \frac{\{c - E(c)\}^2}{E(c)} + \frac{\{d - E(d)\}^2}{E(d)} \quad (5.6)$$

but for the 2×2 table it can be shown that

$$\chi^2_{[1]} = \frac{n(ad - bc)^2}{(a + b)(c + d)(a + c)(b + d)} \quad (5.7)$$

which enables us to calculate χ^2 directly from observed values without the necessity of computing expected values. However, the idea of expected values is essential for the discussion that follows.

POSITIVE AND NEGATIVE ASSOCIATION

In the first method of calculation, Equation 5.6, if $a > E(a)$ this means that there are more occurrences of both species together than would be expected on the basis of no association; and if χ^2 is large enough it may reflect a real positive association between the two species. This implies that the two species tend to occur together. Conversely, if $a < E(a)$ there are fewer occurrences of both species than would be expected on the basis of no association, and again if χ^2 were sufficiently large it may be indicative of a real negative association between the two species, implying that the two species tend not to occur together. Note, however, that χ^2 is always positive because it is a squared quantity. The magnitude of χ^2 is a measure of the degree of association, but the kind of association is given by the sign of $\{a - E(a)\}$.

In the second method of calculation, Equation 5.7, if the product ad is greater than the product bc a positive association is indicated, and conversely if $ad < bc$ the association is negative. Thus the kind of association is given by the sign of $(ad - bc)$, but again χ^2 itself is always positive since this bracketed term is squared in Equation 5.7.

THE MAGNITUDE AND SIGNIFICANCE OF ASSOCIATION

As stated above, χ^2 calculated from a contingency table is a measure of the magnitude of association between two species for either positive or negative associations; the higher the value of χ^2 the higher the degree of association, but comparisons of this kind can only be made between data sets having the same number of stands since χ^2 also increases with sample number. But χ^2 is best known for its use as a test statistic in a statistical test of significance, and can be used as such in the present context.

Employing χ^2 to test the *significance* of association between two species involves exactly the same computations as above, and the only additional action required is to compare the calculated value of χ^2 with the tabulated value for the relevant degrees of freedom (1 for a 2×2 contingency table) and selected probability level (see any standard statistical text for the method). The null hypothesis is that there is no association between the two species, and if the calculated value of χ^2 is equal to or greater than the tabulated value at a particular probability level then the null hypothesis is rejected at that probability level. We then say that χ^2 is *significant* at that probability, and this implies that association is not zero, that is, there is some correlation of occurrence between the two species.

We thus have two separate, though related, uses of χ^2 and it is important to draw the distinction between them. On the one hand, χ^2 simply measures the degree of association; the higher the value of χ^2 the greater the degree of correlation of occurrence of a pair of species in a given number of stands. On the other hand, χ^2 is used as a test statistic to assess the statistical significance of correlation; χ^2 is testing the hypothesis that there is no correlation of occurrence between the two species. Here, the number of stands is irrelevant except that, as in any other significance test, sensitivity is lost when stand number is low; in other words, when stand number is low rejection of the null hypothesis is only possible if the level of association is extreme.

In the method of classification of survey data which relies on χ^2 as a measure of association (Association Analysis, p. 98), it is used as such, and no statistical test is involved. However, at the more detailed level of investigations χ^2 is normally used to assess the significance of association.

χ^2 AS A CORRELATION COEFFICIENT

Random variables, or **variates,** are of two kinds – continuous and discrete. **Continuous variates** can assume any number in a given range, and an example of a pair of such variates is leg length and arm length in humans. If a measure of correlation is required between such variates for a group of people, then the product-moment, or Pearson's correlation coefficient is calculated from the formula

$$r = \frac{\sum\limits_{i=1}^{n} x_i y_i - \left(\sum\limits_{i=1}^{n} x_i \sum\limits_{i=1}^{n} y_i\right)\Big/ n}{\sqrt{\left[\left\{\sum\limits_{i=1}^{n} x_i^2 - \left(\sum\limits_{i=1}^{n} x_i\right)^2\Big/ n\right\}\left\{\sum\limits_{i=1}^{n} y_i^2 - \left(\sum\limits_{i=1}^{n} y_i\right)^2\Big/ n\right\}\right]}}$$

where x_i and y_i are values of each variate for the ith person (or, more generally, the ith observation), and there are n observations. The value of r ranges from -1 (perfect negative correlation) through 0 (no correlation) to 1 (perfect positive correlation), and forms a very useful measure of

76

correlation because r lies in a definite range independent of such things as the number of items in the data.

Numbers of stands containing certain plant species are discrete variates—only whole number values can be assumed. It can be shown that the quantity $\sqrt{(\chi^2/n)}$ has the numerical properties enumerated in the previous paragraph, and so can be applied as a measure of species association which is independent of stand number: we have

$$r = \frac{ad - bc}{\sqrt{\{(a + b)(c + d)(a + c)(b + d)\}}} \qquad (5.8)$$

and the sign of r is the same as that of $(ad - bc)$, as previously discussed.

The product-moment correlation coefficient is a perfectly good *measure* of correlation whatever the statistical distribution of the bivariate data might be. However, for the product-moment correlation coefficient to be used in a significance test, the data must be approximately bivariate normally distributed. Note that the quantity $r = \sqrt{(\chi^2/n)}$ is not tested for significance directly: χ^2 is tested, as described above.

We now present a few examples to illustrate the methods in practice.

Example 5.1

In the Artificial Data set (p. 44), we will examine the associations between Species I and II, and between Species III and IV.

For I and II, 4 stands contain both species (a), 1 stand contains I but not II (b), 2 stands contain II but not I (c), and 3 stands contain neither species (d). So the contingency table is

		Species I		
		+	−	
Species	+	4 (3)	2	6
II	−	1	3	4
		5	5	10

This table checks against the data table, both showing the total occurrences of Species I to be 5, and 6 occurrences of Species II; moreover, adding either the marginal totals at the bottom, or those on the right-hand side,

77

yields the total number of stands involved, 10. Thus, from Equation 5.7:

$$\chi^2_{[1]} = \frac{10\{(4)(3) - (1)(2)\}^2}{(5)(5)(6)(4)} = \frac{(10)(100)}{(25)(24)} = 1.67$$

This represents a positive association, but is not significant. In fact, $E(a) = (5)(6)/10 = 3$; so only one more stand contains both species than would be expected on the basis of no association. Further,

$$r = \sqrt{(\chi^2/n)} = \sqrt{(1.67/10)} = 0.41$$

For Species III and IV, the contingency table is

		Species III		
		+	−	
Species IV	+	0 (1.5)	3	3
	−	5	2	7
		5	5	10

and

$$\chi^2_{[1]} = \frac{10\{(0)(2) - (5)(3)\}^2}{(5)(5)(3)(7)} = \frac{(10)(-15)^2}{(25)(21)} = 4.29^*$$

which is significant at the 5% level of probability, and represents a negative association. The correlation coefficient may be calculated as

$$r = \frac{-15}{\sqrt{\{(5)(5)(3)(7)\}}} = -0.66^*$$

Even though in this data set the species could not be more negatively associated (in the sense that no stand contains both species) the result is only significant at 5% and the correlation coefficient is nowhere near -1. This is a reflection of having 2 stands containing neither species. If each species occurred in 5 stands, but again no stand containing both species, we would have

$$\chi^2 = \frac{10(0 - 25)^2}{(5)^4} = 10.00^{**}$$

78

and $r = -1^{**}$. Even here, the result is only significant at 1% probability because of the small number of stands involved.

The asterisk codes used throughout this book are standard in statistical tests of significance. One asterisk denotes significance at $P(0.05)$, 5% probability, 'significant'. Two asterisks denote significance at $P(0.01)$, 1% probability, 'highly significant'. Three asterisks denote significance at $P(0.001)$, 0.1% probability, 'very highly significant'.

Example 5.2

In woodland, the herb *Mercurialis perennis* occurs on base-rich soils, whereas *Hyacinthoides non-scripta* and especially *Oxalis acetosella* occur on more acid soils. In Coed Nant Lolwyn (Case Study no. 2) the associations between *Mercurialis perennis* and each of the other two species are now investigated.

The contingency table for *Mercurialis perennis* and *Hyacinthoides non-scripta* is

		Mercurialis perennis		
		+	−	
Hyacinthoides non-scripta	+	32 (34.4)	53	85
	−	49	66	115
		81	119	200

the number in brackets indicating the expected number of stands on the basis of no association, $E(a)$. The calculation is

$$\chi^2_{[1]} = \frac{200\{(32)(66) - (53)(49)\}^2}{(81)(119)(85)(115)} = \frac{200(-485)^2}{94\,221\,225} = 0.50$$

and $r = -0.05$. It is obvious that the slight negative association is quite insignificant. Hence, in Coed Nant Lolwyn there is no evidence to suggest any correlation of occurrence between these two species.

79

For *Mercurialis perennis* and *Oxalis acetosella* the contingency table is

		Mercurialis perennis		
		+	−	
Oxalis acetosella	+	11 (28.4)	59	70
	−	70	60	130
		81	119	200

and the calculation is

$$\chi^2_{[1]} = \frac{200\{(11)(60) - (70)(59)\}^2}{(81)(119)(70)(130)} = \frac{200(-3470)^2}{87\ 714\ 900} = 27.46^{***}$$

and $r = -0.37^{***}$, which shows a very highly significant ($P < 0.001$) negative association between these two species.

As an example of a significant positive association, we take *Calluna vulgaris* and *Erica cinerea* from the Iping Common data (Case Study no. 1). The contingency table is

		Calluna vulgaris		
		+	−	
Erica cinerea	+	25 (16)	3	28
	−	15	27	42
		40	30	70

$$\chi^2_{[1]} = \frac{70\{(25)(27) - (15)(3)\}^2}{(40)(30)(28)(42)} = \frac{70(630)^2}{1\ 411\ 200} = 19.69^{***}$$

and $r = 0.53^{***}$. Note that the significance test is made on χ^2, not on r directly.

EFFECT OF QUADRAT SIZE ON ASSOCIATION

The size of sampling unit has a profound effect on the result of association between species, as the following artificial example shows (Fig. 5.1).

In the area there are three plant species: A, B and C. A and B obviously have some degree of association, while C appears uncorrelated with either A or B. Now consider sampling with three different quadrat sizes. Size 1 is roughly the same area as occupied by an individual plant; it is rarely possible for a placing of this size of quadrat to contain more than one species, even in the densely occupied areas of closely associated species A and B. Thus even here, fewer quadrats containing both these species will be found than expected even assuming no association between A and B. There will be a strong negative association *indicated* between all three species, purely as a result of spatial exclusion – that is, the quadrat is not big enough to expose the true situations that exist.

Quadrat size 2 would show Species A and B to have a positive association, and C would be unassociated with the other two. This is the true situation, and is reflected in the data acquired because the area of Quadrat 2 is approximately equal to the area of the clusters of highly associated A and B. With Quadrat 3, every placing would contain all three species, and so no information on any underlying associations could be conveyed: indeed, χ^2 would incalculable.

Evidently, differing sizes of sampling unit can give totally different results for a single situation. The picture is even more complicated when the species differ greatly in size. If at all possible, it would be highly desirable to use

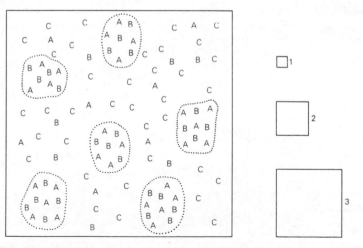

Figure 5.1 The dependence of apparent association between two species on quadrat size (after Kershaw & Looney 1985, by permission of Edward Arnold).

several different quadrat sizes in your study area, and examine the results particularly with regard to the morphology of the species of interest.

Chi-square as a measure and test statistic of stand similarity

ASSOCIATION BETWEEN STANDS

In the context of stand similarity, χ^2 is less often used than other measures. However, it is instructive to compare this use of χ^2 with its employment as a measure of species association, and there is one classification method where χ^2 is explicitly used as the similarity measure.

We are now completely reversing the roles of species and stands. Whereas before, in discussing associations between a pair of species by counting the number of stands in which both species occur, the number of stands in which each of the two species occurs singly, and the number of stands containing neither species; we now examine the similarity between two stands by counting the number of species they have in common, the number of species occurring in either one or the other stand, and the number of species absent from both stands but occurring elsewhere in the data set. The contingency table appears as in Table 5.2. The + + cell in the table, a, now represents the number of species occurring in both stands, b is the number of species occurring in Stand 1 but not Stand 2, c is the number in Stand 2 but not Stand 1, and d is the number of species in the data set as a whole but occurring in neither Stand 1 nor 2. Apart from this reversal of roles, the procedure and interpretation is exactly the same as in the species association case.

Table 5.2 A 2×2 contingency table for assessing association between two stands, 1 and 2.

		Stand 1		
		+	−	
Stand 2	+	a $E(a)$	c $E(c)$	$a + c$
	−	b $E(b)$	d $E(d)$	$b + d$
		$a + b$	$c + d$	$n = a + b + c + d$

Example 5.3

For Stands 4 and 5 in the Artificial Data table (p. 45), we have the following contingency table:

		Stand 4		
		+	−	
Stand 5	+	2 (1.2)	0	2
	−	1	2	3
		3	2	5

$$\chi^2_{[1]} = \frac{5\{(2)(2) - (1)(0)\}^2}{(3)(2)(2)(3)} = \frac{5(4)^2}{36} = 2.22$$

and $r = 0.67$. A degree of positive association exists, but it is non-significant.

Now examine Stands 2 and 3.

		Stand 2		
		+	−	
Stand 3	+	0 (0.8)	2	2
	−	2	1	3
		2	3	5

$$\chi^2_{[1]} = \frac{5\{(0)(1) - (2)(2)\}^2}{(2)(3)(2)(3)} = 2.22$$

and $r = -0.67$, indicating a non-significant negative association.

The precise meaning of a positive association between a pair of stands is that their species contents are relatively similar − they have a large number of species in common, and probably come from a similar vegetation type. Conversely, two stands with a high negative association have very few species in common; so they are dissimilar in species make-up and probably come from different vegetation types.

SIMILARITY BETWEEN STANDS (χ^2)

Neither the χ^2 nor the r can be considered as a measure of similarity directly, although they can be interpreted as measures of association between the two stands analogously to the species situation. With two stands containing 3 and 2 species, respectively, the most similar situation would be

		Stand A		
		+	−	
Stand	+	3	0	3
B	−	0	2	2
		3	2	5

giving

$$\chi^2_{[1]} = \frac{5(6-0)^2}{36} = 5^*$$

and $r = 1^*$; and the most dissimilar situation would be

		Stand A		
		+	−	
Stand	+	0	3	3
B	−	2	0	2
		2	3	5

giving

$$\chi^2_{[1]} = \frac{5(0-6)^2}{36} = 5^*$$

and $r = -1^*$.

Now a measure of similarity should be zero for two entirely different stands, to a high positive value for two identical stands. From the above two examples, we infer that χ^2 is equal to the total number of species in the data set, m, for both completely similar and dissimilar stands. If we artificially attach a negative sign to χ^2 for negative associations, then the total possible range of χ^2 is $m - (-m) = 2m$. Writing χ_s^2 for χ^2 with attached sign, a measure of similarity is given by

$$S_\chi = m + \chi_s^2$$

So for the identical stands above, $S_\chi = 5 + 5 = 10$, and for the completely dissimilar stands $S_\chi = 5 + (-5) = 0$. For Stands 4 and 5 in Example 5.3, we have $S_\chi = 5 + 2.22 = 7.22$, and Stands 2 and 3 in Example 5.3 yield $S_\chi = 5 + (-2.22) = 2.78$. In the situation of no association between two stands, χ^2 is theoretically zero giving $S = m$, which is in the middle range of S_χ. This is as it should be, because such a pair of stands would have some species in common purely by chance.

Note that it is impossible to test the significance of similarity; S_χ is simply a measure of similarity of two stands in a data set of m species, which varies between 0 (totally dissimilar) and $2m$ (identical).

SIMILARITY BETWEEN STANDS (r)

A better measure of stand similarity is based on the correlation coefficient because its range of values is independent of species number. Since for perfect similarity $r = 1$, and for extreme dissimilarity $r = -1$, the total range is 2; and the similarity coefficient is given by

$$S_r = 1 + r$$

which is zero for complete dissimilarity and 2 for identical stands. On this basis, Stands 4 and 5 in Example 5.3 give 1.67, and Stands 2 and 3 have a similarity coefficient of 0.33.

Similarity measures are more immediately interpretable if they lie in the range 0 to 1, from which they are directly convertible to percentages. This is easily achieved for the χ^2 similarity coefficient by dividing by $2m$, and for the r coefficient by dividing by 2. In what follows, all similarities will be expressed on a 0 to 1 scale.

Example 5.4

In the Iping Common data we examine the associations and similarities between Stands 0 and 14, and Stands 0 and 62.

For Stands 0 and 14, the contingency table is

		Stand 0		
		+	−	
Stand 14	+	5 (1.9)	1	6
	−	3	17	20
		8	18	26

$$\chi^2_{[1]} = \frac{26\{(5)(17) - (3)(1)\}^2}{(8)(18)(6)(20)} = \frac{26(82)^2}{17\,280} = 10.12^{**},$$

$r = 0.62^{**}$, and $S_r = 0.81$. The two stands are in the same vegetation type, with a dominance of *Calluna vulgaris* and a presence of several lichen species.

For Stands 0 and 62, we have

		Stand 0		
		+	−	
Stand 62	+	0 (0.6)	2	2
	−	8	16	24
		8	18	26

$$\chi^2_{[1]} = \frac{26\{(0)(16) - (8)(2)\}^2}{(8)(18)(2)(24)} = \frac{26(-16)^2}{6912} = 0.96,$$

$r = -0.19$ and $S_r = 0.41$. Although these two stands have no species in common and come from quite different species assemblages (Stand 62 is dominated by tall *Pteridium aquilinum* with an understorey of *Molinia caerulea*) χ^2 is very small because of the low numbers of species actually occurring in each stand, especially in Stand 62. Typically, any one stand contains only a small number of the species from the complete list and so χ^2, and hence r, S_x, and S_r are rather insensitive measures of stand similarity.

DISSIMILARITY BETWEEN STANDS

For some classification and ordination purposes, a coefficient of dissimilarity between stands is required rather than similarity. Now we need a value of zero for identical stands and some maximum positive value for complete dissimilarity. For the χ^2 we have

$$D_x = 2m - S_x = m - \chi_s^2$$

and for the correlation coefficient

$$D_r = 2 - S_r = 1 - r$$

The two measures, S and D, have the same range and the same end-point values, but opposite meanings.

Other measures of stand similarity

The main disadvantage of χ^2 (or r) as a measure of stand similarity is highlighted by Stands 0 and 62 in Example 5.4. Although the two stands had no species in common, there was no significant negative association between them, and their similarity measure was not particularly low either ($S_r = 0.41$). This is because χ^2 measures association between the two stands, not just on the basis of the number of species they have in common, but also taking account of the number of species that both stands *do not* have in common. In samples of normal vegetation this is usually quite large, and so it is rare to obtain low similarity values of two stands based on χ^2 even though the vegetation types appear totally different. One way out of this difficulty is to use a measure which is based on species presences only; in the notation of Table 5.2, a coefficient which is calculated from a, b, and c, but not d. Two related coefficients are commonly used which have ranges of 0 (complete dissimilarity) to 1 (identical).

JACCARD'S COEFFICIENT

This is given by

$$S_J = \frac{a}{a + b + c} \tag{5.9}$$

for similarity, and

$$D_J = \frac{b + c}{a + b + c} \tag{5.10}$$

for dissimilarity.

SORENSEN'S COEFFICIENT

This coefficient is very similar to the last, and is given by

$$S_S = \frac{2a}{2a + b + c} \qquad (5.11)$$

for similarity, and

$$D_S = \frac{b + c}{2a + b + c} \qquad (5.12)$$

for dissimilarity.

Example 5.5

In Stands 0 and 62 of the Iping Common data, both the Jaccard and Sorensen coefficients are zero (because $a = 0$), which is a sensible result for two stands having no species in common; remember that $S_r = 0.41$. We now investigate, from the same data set, Stands 0 and 70. The contingency table is

		Stand 0		
		+	−	
	+	2 (1.9)	4	6
Stand 70				
	−	6	14	20
		8	18	26

The expected number of species common to both stands, 1.9, on the basis of no association is very close to the observed number, 2; χ^2 will evidently be very small, even though the two stands come from clearly different vegetation types (see pp. 50–2 and Figs 4.4 & 4.5). We have

$$\chi^2 = \frac{26(28 - 24)^2}{(8)(18)(6)(20)} = \frac{26(4)^2}{17\,280} = 0.02$$

Moreover, the association is actually positive! Hence, $r = 0.03$ and so $S_r = 0.52$. For our other coefficients, we have

$$S_J = \frac{2}{2 + 6 + 4} = \frac{2}{12} = 0.17$$

and

$$S_S = \frac{4}{4 + 6 + 4} = \frac{4}{14} = 0.29$$

Intuitively, for this pair of stands, Jaccard's coefficient would appear to be the best similarity measure.

Quantitative data

The product-moment correlation coefficient

The product-moment correlation coefficient is given by

$$r = \frac{\sum\limits_{j=1}^{m} x_j y_j - \left(\sum\limits_{j=1}^{m} x_j \sum\limits_{j=1}^{m} y_j \right) \Big/ m}{\sqrt{\left[\left\{ \sum\limits_{j=1}^{m} x_j^2 - \left(\sum\limits_{j=1}^{m} x_j \right)^2 \Big/ m \right\} \left\{ \sum\limits_{j=1}^{m} y_j^2 - \left(\sum\limits_{j=1}^{m} y_j \right)^2 \Big/ m \right\} \right]}} \qquad (5.13)$$

where x_j and y_j are the abundance values of the jth species in the first and second stands, respectively, and each summation is over all the m species. As in the case of r derived from χ^2 for qualitative data, r lies in the range -1 through zero to 1, and so indicates whether the stands are positively or negatively correlated as regards species composition and abundance. As before, the corresponding similarity measure is

$$S_r = 1 + r$$

which can be divided by 2 so that S_r lies in the range 0–1.

It is most unwise to test the significance of r because this is only valid if the distribution of species abundances in the two stands are approximately bivariate normal. Owing to the very large number of zeros in most vegetation data sets (see examples below), such distributions are highly anormal.

Spearman's rank correlation coefficient

A rank correlation coefficient, where the significance test does not depend on the data being at least approximately bivariate normally distributed, is appropriate if one wishes to test the hypothesis of no correlation of the abundance of two species. The best known rank correlation coefficient is that of Spearman.

89

We first calculate

$$d = \sum_{j=1}^{m} (x_j - y_j)^2 \qquad (5.14)$$

where x_j and y_j are the *ranked* abundance values of the jth species in the first and second stands, respectively, and the summation is over all m species. In Equation 5.14, d is the sum of squares of the differences of the rankings of the species in the two stands. If there are ties, then each observation of the tied rank is replaced by the average across all the observations of that rank. For example, suppose there are 4 species absent (each score zero. each rank 1), then each of the 4 species would be given the rank 2.5 ($= (1 + 2 + 3 + 4)/4$).

If there are no ties, then the correlation coefficient is given by

$$r_S = 1 - \frac{6d}{m^3 - m} \qquad (5.15)$$

If there are ties, the procedure is more complicated. Suppose there are p ties of x-rankings and q ties of y-rankings; we then calculate

$$T_x = \sum_{i=1}^{p} \frac{t_i^3 - t_i}{12}$$

for ties of x-rankings, and

$$T_y = \sum_{i=1}^{q} \frac{t_i^3 - t_i}{12}$$

for ties of y-rankings, where t_i is the number of items in the ith tie. Then

$$r_S = \frac{X + Y - d}{2\sqrt{(XY)}} \qquad (5.16)$$

where $X = \{(m^3 - m)/12\} - T_x$ and $Y = \{(m^3 - m)/12\} - T_y$. Practical applications of both these correlation coefficients are given in Example 5.7.

Czekanowski's coefficient

The formula for Czekanowski's coefficient is

$$S_C = \frac{2 \sum_{j=1}^{m} \min(x_j, y_j)}{\sum_{j=1}^{m} x_j + \sum_{j=1}^{m} y_j} \qquad (5.17)$$

where $\min(x_j, y_j)$ represents the lesser of the scores of species j in the two stands, and the remaining notation is the same as in Equation 5.13. The range of the coefficient is from 0 (complete dissimilarity) to 1 (identity).

Example 5.6

We shall employ the quantitative version of the Artificial Data to compute the product-moment and Czekanowski coefficients for the same two comparisons as in Example 5.3, i.e. Stands 4 and 5, and Stands 2 and 3.

For Stands 4 and 5 we have: $\Sigma x = 5$, $\Sigma x^2 = 9$ (Stand 4); $\Sigma y = 10$, $\Sigma y^2 = 58$ (Stand 5); $m = 5$, $\Sigma xy = 21$. Substituting in Equation 5.13 gives

$$r_{[3]} = \frac{11}{\sqrt{\{(4)(38)\}}} = 0.89^*$$

and $S_r = 0.95$.

For the Czekanowski coefficient,

$$\Sigma \min(x_j, y_j) = 0 + 2 + 0 + 2 + 0 = 4$$

There is one term in the summation per species, and each term is the lesser abundance value for that species. So for Species I, Stand 5 has the lower score (0); for Species II, Stand 4 has the lower score (2), and so on. So

$$S_C = \frac{(2)(4)}{5 + 10} = \frac{8}{15} = 0.53$$

For Stands 2 and 3 the same calculations yield: $\Sigma x = 11$, $\Sigma x^2 = 101$ (Stand 2); $\Sigma y = 11$, $\Sigma y^2 = 101$ (Stand 3); $m = 5$, $\Sigma xy = 0$;

$$r_{[3]} = \frac{-24.2}{\sqrt{\{(76.8)(76.8)\}}} = -0.32$$

and $S_r = 0.34$, but $S_C = 0$ because there are no species in common; so the lesser of each species is always zero.

Example 5.7

In the Iping Common data, we have compared Stands 0 & 14, 0 & 62, and 0 & 70 on a qualitative basis. We now compare them from a cover abundance viewpoint.

If you are doing such calculations by hand, rather than using a more comprehensive computer program, the data need to be set out rigorously but economically, as shown in Table 5.3 for Stands 0 and 14. These data give $r_{[25]} = 0.97$*** and $S_r = 0.99$. For the Czekanowski coefficient

$$S_C = \frac{(2)(89)}{115 + 97} = 0.84$$

For Spearman's correlation coefficient the ranking scores are listed in Table 5.4. So that you can see exactly what is going on, it would be advisable for you to write out the whole 2×26 data matrix. The calculations are then:

$d = 474$ (see Table 5.5)

$T_x = [\{(18)^3 - 18\} + \{(2)^3 - 2\} + \{(2)^3 - 2)\} + \{(2)^3 - 2\}]/12 = 486$

$T_y = [\{(20)^3 - 20\} + \{(2)^3 - 2\}]/12 = 665.5$

$X = \{(26)^3 - 26\}/12 - 486 = 976.5$

$Y = \{(26)^3 - 26\}/12 - 665.5 = 797$

Table 5.3 Arrangement of data for calculation of similarity coefficients for Stands 0 and 14 of the Iping Common data.

Species	Abundance of species		x^2	y^2	xy	$\min(x_j, y_j)$
	Stand 0 (x)	Stand 14 (y)				
Cladonia coccifera	5	2	25	4	10	2
Cladonia crispata	5	2	25	4	10	2
Cladonia coniocraea	2	0	4	0	0	0
Cladonia floerkiana	2	0	4	0	0	0
Hypogymnia physodes	1	0	1	0	0	0
Campylopus interoflexus	20	20	400	400	400	20
Polytrichum juniperinum	0	3	0	9	0	0
Calluna vulgaris	60	65	3600	4225	3900	60
Erica cinerea	20	5	400	25	100	5
17 others	0	0	0	0	0	0
$m = 26$						
$\Sigma x, \Sigma y$	115	97				
$\Sigma x^2, \Sigma y^2$			4459	4667		
Σxy					4420	
$\Sigma \min(x_j, y_j)$						89

Table 5.4 Ranking scores for data given in Table 5.3.

Abundance of species in the stand	Number of species of this abundance	Ranking score
Stand 0		
0	18	9.5
1	1	19
2	2	20.5
5	2	22.5
20	2	24.5
60	1	26
Stand 14		
0	20	10.5
2	2	21.5
3	1	23
5	1	24
20	1	25
65	1	26

Table 5.5 Calculation of d (the sum of squares of the differences of the rankings of the species in the two stands, see Eqn 5.14).

Species	Ranked abundances of species		
	Stand 0 (x_j)	Stand 14 (y_j)	$(x_j - y_j)^2$
Cladonia coccifera	22.5	21.5	1.0
Cladonia crispata	22.5	21.5	1.0
Cladonia coniocraea	20.5	10.5	100.0
Cladonia floerkiana	20.5	10.5	100.0
Hypogymnia physodes	19	10.5	72.25
Campylopus interoflexus	24.5	25	0.25
Polytrichum juniperinum	9.5	23	182.25
Calluna vulgaris	26	26	0
Erica cinerea	24.5	24	0.25
17 others	9.5	10.5	1
$d = \Sigma(x_j - y_j)^2$			474

Table 5.6 Data for calculation of similarity coefficients for Stands 0 and 62 of the Iping Common data.

Species	Abundance of species		x^2	y^2	xy	$\min(x_j, y_j)$
	Stand 0 (x)	Stand 62 (y)				
Cladonia coccifera	5	0	25	0	0	0
Cladonia crispata	5	0	25	0	0	0
Cladonia coniocraea	2	0	4	0	0	0
Cladonia floerkiana	2	0	4	0	0	0
Hypogymnia physodes	1	0	1	0	0	0
Campylopus interoflexus	20	0	400	0	0	0
Calluna vulgaris	60	0	3600	0	0	0
Erica cinerea	20	0	400	0	0	0
Molinia caerulea	0	75	0	5625	0	0
Pteridum aquilinum	0	80	0	6400	0	0
16 others	0	0	0	0	0	0
$m = 26$						
Totals (Σ)	115	155	4459	12 025	0	0

Table 5.7 Data for calculation of similarity coefficients for Stands 0 and 70 of the Iping Common data.

Species	Abundance of species		x^2	y^2	xy	$\min(x_j, y_j)$
	Stand 0 (x)	Stand 70 (y)				
Cladonia coccifera	5	0	25	0	0	0
Cladonia crispata	5	0	25	0	0	0
Cladonia coniocraea	2	0	4	0	0	0
Cladonia floerkiana	2	0	4	0	0	0
Hypogymnia physodes	1	0	1	0	0	0
Ulex minor	0	5	0	25	0	0
Campylopus interoflexus	20	0	400	0	0	0
Calluna vulgaris	60	40	3600	1600	2400	40
Erica cinerea	20	10	400	100	200	10
Erica tetralix	0	20	0	400	0	0
Molinia caerulea	0	10	0	100	0	0
Pteridum aquilinum	0	2	0	4	0	0
14 others	0	0	0	0	0	0
$m = 26$						
Totals (Σ)	115	87	4459	2229	2600	50

with 24 degrees of freedom.

$$r_S = \frac{976.5 + 797 - 474}{2\sqrt{(976.5)(797)}} = 0.74^{***}$$

The data for Stands 0 and 62 are given in Table 5.6. They give $r_{[25]} = -0.10$, $S_r = 0.45$ and $S_C = 0$.

Data for Stands 0 and 70 (Table 5.7) give $r_{[25]} = 0.80^{***}$, $S_r = 0.9$ and

$$S_C = \frac{(2)(50)}{115 + 87} = \frac{100}{202} = 0.50$$

A comparative discussion on the magnitudes of these coefficients is given in the next section.

Comparisons of similarity coefficients

The measures of similarity discussed in this chapter by no means exhaust the long list of coefficients that have been used in vegetation analysis over the years, but they are sufficient in an introductory study. Greig-Smith (1983) provides a comprehensive list, together with a consideration of their properties.

Table 5.8 draws together the results of all worked examples on similarity coefficients in this chapter, using the five examples (two from the Artificial Data and three examples of real data), and five coefficients: χ^2 correlation, Jaccard, and Sorensen coefficients for qualitative data; product-moment correlation, and Czekanowski coefficients for quantitative data. Examination of this small set of examples leads to the following observations:

(a) Quantitative measures are usually higher than the corresponding qualitative coefficients for the same data set. The only exception in Table 5.8 is that the Czekanowski coefficient is less than both the Jaccard and Sorensen coefficients in Stands 4 and 5 of the Artificial Data. An explanation of this anomaly may lie in the fact that overall species abundances are twice as great in Stand 5 as they are in Stand 4. If the abundances in Stand 4 are doubled, then the total abundances become the same in both stands and Czekanowski's coefficient is now 0.7, which puts it in the same range as the corresponding qualitative coefficients of Jaccard and Sorensen.

(b) The correlation coefficients, both qualitative and quantitative, are always higher than the other coefficients in each data category. The worst effect of this propensity is the assignment of a non-zero similarity

Table 5.8 Comparisons of similarity coefficients, for both qualitative and quantitative data. S_r – derived from a correlation coefficient, and expressed on the 0–1 scale; S_J – Jaccard's coefficient; S_S – Sorensen's coefficient; S_C – Czekanowski's coefficient.

Data set and stand numbers	Qualitative			Quantitative		Notes
	S_r	S_J	S_S	S_r	S_C	
Artificial data 4 and 5	0.84	0.67	0.80	0.95	0.53	obviously considerable similarity on species presence, but a large difference in abundance of one pair of common species
Artificial data 2 and 3	0.17	0	0	0.34	0	no species in common, but one species absent from both stands; thus $S_r \neq 0$
Iping Common 0 and 14	0.81	0.56	0.71	0.99	0.84	from a visual point of view, the two stands are from the same vegetation type; but closer examination shows Stand 0 to have 5 lichen species, with total cover of 10%, but Stand 14 has only 2 lichen species with total cover of 2%
Iping Common 0 and 62	0.41	0	0	0.45	0	no species in common, but 16 out of all 26 species in the data set are absent from both stands; hence the relatively high values of S_r
Iping Common 0 and 70	0.52	0.17	0.29	0.90	0.50	stands are from two different vegetation types, but there are two species in common, and one of these (*Calluna*) has high cover in both stands

value by a correlation coefficient to a pair of stands having no species in common (Artificial Data, Stands 2 and 3; Iping Common, Stands 0 and 62). The latter data give a ridiculously high qualitative correlation coefficient for two stands having no species in common. The reason for this is that a large proportion of the total species number in the entire data set (16 out of 26, or nearly two-thirds) is absent from both stands, and it is in this respect that the correlation coefficient 'regards' the stands as similar. Jaccard's, Sorensen's, and Czekanowski's coefficients, on the other hand, take no cognizance of species which are absent in both stands.

You may very reasonably ask why correlation coefficients are ever used in plant ecological work when they seem to have such undesirable properties. The answer probably lies in the fact that the product-moment correlation coefficient and χ^2 are standard quantities in statistics and have been generally applied to innumerable situations. Moreover, their statistical distributions are known, and so they can be employed in tests of hypotheses. The other coefficients have been derived and used in ecology only, and their statistical distributions are mostly unknown. The choice between these two kinds of similarity measures rests upon the importance of common absent species: since these usually dominate a pair of stands, correlation coefficients will be heavily influenced by these species. More usually, however, similarity between two stands is judged by species presences rather than absences, and so the non correlation coefficients seem to be ecologically superior.

6

Classification

This chapter, and the next on ordination, examines some of the methods that have been used to summarize data from vegetation surveys involving many species. The broad underlying principles of, and distinctions between, classification and ordination have already been discussed in Chapter 3. These two chapters are concerned with methodological details and general comparisons among classificatory and ordination methods.

To start, we shall survey one of the oldest techniques to classify stands – normal Association Analysis. Although the method in its strict form is now obsolete, it is worth describing in detail because: (a) it is conceptually simple; (b) it is still useful in modified form; (c) it uses the χ^2 measure of species association which was fully discussed in the previous chapter; (d) it is simple to interpret; and (e) it forms a basis for us to view other classification methods.

Normal Association Analysis

Association Analysis was first described by Williams & Lambert (1959, 1960) but the method is similar to an older one due to Goodall (1953). Starting with a total of *n* stands, the aim is to *progressively* subdivide them into a small number of final groups. At each stage, division has to be accomplished on the basis of some criterion, and we also need some other criterion to stop further subdivision at a certain point.

Starting with the data matrix, χ^2 is calculated for all possible species pairs by Equation 5.7; in terms of the data matrix we calculate a value of χ^2 for all possible pairs of columns. The total number of χ^2-values is $^mC_2$, the combination of *m* species or columns taken two at a time, and is given by

$$^mC_2 = \frac{m!}{2!(m-2)!} = \frac{m(m-1)}{2}$$

Two examples of the calculation have already been given using the Artificial Data (pp. 77–82), for Species I & II where the result is 1.67, and Species III & IV which gives $\chi^2 = 4.29$. We are unconcerned here whether χ^2 indicates positive or negative associations, neither are we interested if it departs significantly from zero; only the magnitude of χ^2 matters.

The $m(m-1)/2$ chi-square values are best arranged in an $m \times m$ matrix, where both the rows and the columns correspond to species. Then either the rows or the columns of the χ^2-matrix are summed, and the results are sum-chi-squares, $\Sigma\chi^2$, for each species. The species with the highest $\Sigma\chi^2$ is designated the divisor species, and the group of n stands is divided into two sub-groups on the basis of whether a stand does or does not contain the divisor species. Then the whole process is repeated on the sub-groups. Example 6.1 shows the detailed working of the method using the Artificial Data set.

Example 6.1

Using all 10 stands of the Artificial Data, containing the 5 species, $(5)(4)/2 = 10$ χ^2-values are calculated and set out in a 5×5 matrix.

	I	II	III	IV	V	$\Sigma\chi^2$
I	—	1.67	0.40	0.48	1.11	3.67
II	1.67	—	1.67	0.08	1.67	5.09
III	0.40	1.67	—	4.29	1.11	$7.47 = (\Sigma\chi^2)_{max}$
IV	0.48	0.08	4.29	—	0.48	5.33
V	1.11	1.67	1.11	0.48	—	4.37

The matrix is symmetric, so only the elements on one side of the leading diagonal need be quoted; in future, only the elements beneath and to the left of the leading diagonal will be displayed for symmetric matrices.

Species III has the highest sum-chi-square, and so it is the divisor species; consequently, the stands are divided into two groups: 'III + ', those stands which contain Species III, namely Stands 1, 3, 6, 8, 10; and 'III − ', those stands not containing Species III, i.e. Stands 2, 4, 5, 7, 9. Thus the first division has been into two equal groups.

We now repeat the process on each of the sub-groups. First, we display the data matrix for the stands of group III + :

	I	II	V
1	1	0	0
3	0	0	1
6	1	1	0
8	0	0	0
10	0	1	0

99

The column for Species III has been omitted because it now occurs in every stand of the sub-group; further, Species IV does not occur in any of the stands of group III + , and so this column is also deleted. Chi-square cannot be calculated for a species which is either present in, or absent from, all stands; using the formula, one obtains $\chi^2 = 0/0$, which is indeterminate. There are now only $(3)(2)/2 = 3$ χ^2-values to compute, and these are:

	I	II	V	$\Sigma\chi^2$
I	—			0.97
II	0.14	—		0.97
V	0.83	0.83	—	$1.66 = (\Sigma\chi^2)_{max}$

Now Species V is the divisor species, which produces the unequal groupings:

'III + , V + ', i.e. Stand 3; and 'III + , V − ', i.e. Stands 1, 6, 8, 10

Finally, we consider the 'III − ' group of stands:

	I	II	IV
2	1	1	0
4	1	1	1
5	0	1	1
7	0	0	1
9	1	1	0

Here, both Species III and V are missing from all five stands, the former by definition. The χ^2-matrix is

	I	II	IV	$\Sigma\chi^2$
I	—			$4.10 = (\Sigma\chi^2)_{max}$
II	1.88	—		2.71
IV	2.22	0.83	—	3.05

Species I is the divisor species, which yields the two groups:

'III − , I + ', i.e. Stands 2, 4, 9; and 'III − , I − ', i.e. Stands 5, 7

The whole process is usually displayed in diagrammatic form, as in Figure 6.1.

To show what has been accomplished, the whole data matrix is restated in Table 6.1, but rearranged to show the groupings (associations). A better

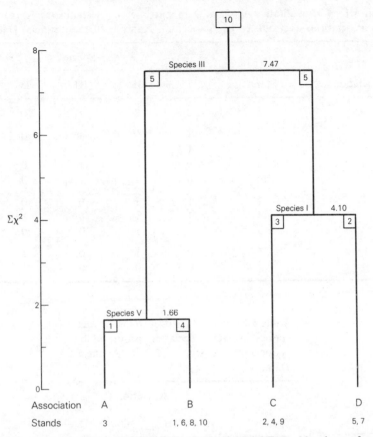

Figure 6.1 Normal Association Analysis of the Artificial Data. Numbers of stands at each stage are shown in boxes, and the divisor species and $(\Sigma\chi^2)_{max}$-values are shown for each division.

way of displaying the data in this fashion will be presented at the end of the chapter. Alternatively, the percentage frequency of each species in each association can be tabulated, as in Table 6.2.

Rationale of χ^2 as the underlying numerical index

The aim of classifying stands of vegetation is to obtain groups of stands of relatively homogeneous composition. One way of defining homo-geneity/heterogeneity in a group of stands is through the amount of correlation of occurrence of the species in those stands. Thus if we have a set of stands in which the species show no significant associations among themselves, in other words the χ^2-values and hence the $\Sigma\chi^2$-values are all

Table 6.1 The qualitative Artificial Data matrix with the stands (rows) rearranged into associations as produced by Association Analysis of the Artificial Data.

| Association | Stand | Species | | | | |
		I	II	III	IV	V
A	3	0	0	1	0	1
B	1	1	0	1	0	0
	6	1	1	1	0	0
	8	0	0	1	0	0
	10	0	1	1	0	0
C	2	1	1	0	0	0
	4	1	1	0	1	0
	9	1	1	0	0	0
D	5	0	1	0	1	0
	7	0	0	0	1	0

Table 6.2 Percentage occurrence of each species in each association produced by Association Analysis of the Artificial Data.

| Species | Association | | | |
	A	B	C	D
I	0	50	100	0
II	0	50	100	50
III	100	100	0	0
IV	0	0	33	100
V	100	0	0	0

low, then we have a homogeneous group of stands. An alternative interpretation of this way of viewing homogeneity is that there are relatively few species occurring with each other in the group of stands. Each homogeneous group, or association, will tend to have a restricted set of species because:

(a) a species (say I) occurring in the group which, in the whole of the surveyed area, has a high positive association with another species (say II) will tend to have the other species occurring with it; and

(b) a species (say III) which has a high negative association with another species (say IV) will tend to 'repel' it.

In other words, if in an association of stands Species I occurs, Species II is likely to be there also; whereas if Species III occurs, Species IV is unlikely to occur. Hence, χ^2 seems an appropriate coefficient to use in this scheme of classification, but other coefficients could be used. Indeed, in some ways χ^2 is not a particularly good measure for this purpose, and its use can lead to misleading results. It is for this reason that Association Analysis has fallen into disfavour.

Levels of division and 'stopping' rules

In Example 6.1 we carried out two levels of division and obtained four stand groups and, as it happens, the groups contain anything from one to four stands. Now, except for Association A which contains only one stand, further levels of division could be achieved; and if the process was continued exhaustively, the end result would be the individual stands and we would be back where we started. Evidently the process has to stop somewhere before this, and the problem is to define where and how. There are two ways of approaching the problem: subjectively, or objectively. In either case, stopping implies drawing a horizontal line across a diagram of the form of Figure 6.1. On moving down from the top of the diagram, divisions become less important; we need to know how far down to draw a line below which further divisions become inconsequential, or excessively fragment the data.

SUBJECTIVE STOPPING

Subjective stopping generally involves deciding oneself how many associations are thought to exist, and then drawing a horizontal line across the diagram. So long as a *horizontal* line is drawn, there is an element of objectivity to the process. For example, let us assume that three associations were deemed appropriate for the Artificial Data rather than the four produced in Example 6.1 and Figure 6.1. The line would need to be drawn in the range $1.66 < \Sigma \chi^2 < 4.10$, thus maintaining Associations C and D, but recombining Associations A and B. By inspection of Figure 6.1 it can be seen that the splitting-off of Stand 3 from Stands 1, 6, 8, 10 is relatively trivial compared with the separation of Stands 5, 7 from 2, 4, 9. So although we may specify the number of associations required, we do not have the same control over their structure.

In Association Analysis, where an objective stopping rule can be applied, it is as well to use it even if we ultimately wish to specify subjectively the number of groups. It is always useful to start with some objectively defined

103

number of associations; then, if there seem to be good ecological reasons for adjusting the number of associations, the program can be re-run with a different stopping level specified.

An objective stopping rule for Association Analysis can be built into the program involving the magnitude of a specified χ^2-value at each potential division (see below). The relevant quantity could be $(\Sigma \chi^2)_{max}$, but the usual one is χ^2_{max} which is the maximum χ^2-value to be found in the chi-square matrix of the potential division being examined. The method works like this.

For a group of stands, a χ^2-matrix is computed: this can be called a *potential division*. If χ^2_{max} in this matrix is greater than or equal to some specified value, then the potential division is actually made in the way already described; if, however, χ^2_{max} is less than the specified value then the division is not made, and the whole group of stands is deemed to be a final group or association.

The remaining problem is to define the level of χ^2_{max}. In the early days of Association Analysis the values of χ^2 for one degree of freedom at $P(0.05)$, $P(0.01)$, or $P(0.001)$ were tried. These stopping levels proved to be too low for most data sets – too many associations were produced. Most vegetation surveys involve at least many tens if not hundreds of stands, and χ^2 increases with stand number for a given degree of species association; indeed, the maximum possible value of χ^2_{max} is equal to the total number of stands in the data set. At Aberystwyth, we have found $\sqrt{n}$, where n is the total number of stands in the data set, to be a useful 'rule of thumb' value of χ^2_{max} in providing an objective stopping level.

Types of classification

Having acquired a detailed knowledge of one particular method of classification, we may now list and briefly comment on the different kinds of classification procedure that are available. There are two main features that define a classification method: the **strategy** – how the arrangements are made which yield the final stand groupings; and the **numerical basis** upon which the strategy works. In this discussion we are concerned only with strategies of classification.

Reticulate and hierarchical classifications

Almost all classification methods used in vegetation analysis are hierarchical. Association Analysis is an example of a hierarchical classification, and the meaning of the term is apparent from the sequence of divisions at different levels of $\Sigma\chi^2$. The final associations of stands are produced from the initial total number of stands by a series of hierarchical divisions.

Reticulate, or non-hierarchical, classifications lack the sequential structure of hierarchical classifications; they have scarcely been used in ecology, and we shall not consider them.

Divisive and agglomerative strategies

In Association Analysis we start with the whole set of stands and progressively subdivide until, if we go the whole way, each stand appears singly at the bottom of the diagram. In other words, the strategy is **divisive**. Many kinds of divisive strategy have been developed. An **agglomerative** strategy starts, so to speak, at the bottom with each stand individually and, by a series of *fusions* of increasing importance, we eventually arrive at the complete set of stands. Again, there are several fusion strategies.

In an agglomerative classification there can be no stopping rule in the strict sense because, as the starting point is the individual stands, the whole hierarchy must be created. A 'stopping' level can then be placed in a secondary fashion as described for Association Analysis, and this line can be placed using subjective or objective criteria.

Monothetic and polythetic criteria

In any hierarchical classification, divisions or fusions at each level are governed by certain criteria. More specifically, a set of objects are classified on the basis of their attributes; in normal Association Analysis, the objects are the stands and the attributes are the species they contain. Association Analysis is an example of a **monothetic** criterion: division of a set of stands at any one level is governed by the presence or absence of a single attribute or species. A strategy in which more than one species determines the division or fusion at each stage is **polythetic**.

The types of hierarchical classification strategy

Combining the two strategies and the two criteria of the previous sections in all possible combinations gives, in theory, four kinds of hierarchical classification: divisive monothetic, divisive polythetic, agglomerative monothetic and agglomerative polythetic. Of these, an agglomerative monothetic strategy can scarcely exist in practice. Consider the fusion of two stands at the beginning of the analysis; with a monothetic criterion there would be a very large number of possible candidate stands for the fusion. Even among the 45 stand pairs of the Artificial Data there are no fewer than 29 equally good candidates for the first fusion based on monothetic criteria (single species) alone.

Few classificatory methods involve complicated mathematics; rather, they involve a vast repetition of elementary operations on a great quantity of numbers. A computer dealing with this kind of situation is said to be 'number crunching'. Nowadays, computer memory capacity is not usually a constraint on the complexity of analysis that can be undertaken on a large data matrix, even many microcomputers can cope; but in the early years, around the beginning of the 1960s, computer size and, to a lesser extent, speed placed a considerable restraint on the ambitions of plant ecologists. Monothetic methods require less in the way of computer resources than polythetic, but it is evident that a polythetic method *should* be superior to a monothetic one. Consequently, divisive monothetic methods were among the earliest kinds of classification to be formulated and used, but early polythetic strategies had to be agglomerative. It is only relatively recently that divisive polythetic methods have become practical propositions.

Comparisons of these three major classification types: divisive monothetic, divisive polythetic, and agglomerative polythetic, will be made after the methods have been described.

Divisive monothetic methods

Association Analysis (Williams & Lambert 1959, 1960)

As this has already been described *in extenso*, it is only necessary to record the method here in its proper place.

Information Statistic (Macnaughton-Smith 1965)

'Information' statistics of various kinds, so-called because of their derivation from information theory in communication engineering, have several applications in vegetation classification. Two differing versions are used in divisive monothetic strategies, hence the inclusion of the author with the title. The Macnaughton-Smith Information Statistic method works in precisely the same way as Association Analysis, the only difference being that an information statistic is calculated from the two-way table (Table 5.1) rather than chi-square. In the notation of Table 5.1, we have:

$$2I_{AB} = 2\{a \log_e a + b \log_e b + c \log_e c + d \log_e d$$
$$- (a+b)\log_e(a+b) - (c+d)\log_e(c+d) - (a+c)\log_e(a+c)$$
$$- (b+d)\log_e(b+d) + n \log_e n\} \tag{6.1}$$

Although rather lengthy, there is a regularity in the succession of terms and it is easy to use. The reason for using $2I$ rather than I is that the numerical values of $2I$, and its probability density function, are very similar to χ^2. Hence the analogy of the two quantities is complete, and the same numerical stopping rule can be used, $\sqrt{n}$.

Example 6.2

The Information Statistic for Species III & IV of the Artificial Data is calculated from the second two-way table in Example 5.1 as

$$2I_{\text{III,IV}} = 2(0.\log_e 0 + 5.\log_e 5 + 3.\log_e 3 + 2.\log_e 2 - 5.\log_e 5 - 5.\log_e 5$$
$$- 3.\log_e 3 - 7.\log_e 7 + 10.\log_e 10)$$
$$= 2(0 + 8.05 + 3.30 + 1.39 - 8.05 - 8.05 - 3.30 - 13.62 + 23.03)$$
$$= 5.50$$

The value of $0.\log_e 0$ is zero.

Example 6.3

Here, the Artificial Data will be employed to illustrate the working of the Macnaughton-Smith Information Statistic classification. Read this example in conjunction with Example 6.1.

Using all 10 stands and 5 species, the $2I$ matrix is

	I	II	III	IV	V	$\Sigma 2I$	
I	—					4.11	
II	1.73	—				5.54	
III	0.40	1.73	—			9.13	$= (\Sigma 2I)_{\max}$
IV	0.48	0.08	5.50	—		6.82	
V	1.50	2.00	1.50	0.76	—	5.76	

Again, as in Association Analysis, Species III has the highest sum-$2I$, and so the first division of the stands is effected on this species as before.

For the 'III + ' group of stands, 1, 3, 6, 8, 10, we obtain the following $2I$

matrix:

	I	II	V	$\Sigma 2I$
I	—			1.33
II	0.14	—		1.33
V	1.19	1.19	—	$2.38 = (\Sigma 2I)_{max}$

Species V is thus the divisor species, as before.

For the 'III − ' group of stands, 2, 4, 5, 7, 9, we get the matrix:

	I	II	IV	$\Sigma 2I$
I	—			$5.14 = (\Sigma 2I)_{max}$
II	2.23	—		3.42
IV	2.91	1.19	—	4.10

so again the result is the same as for Association Analysis, and the corresponding levels of division in the two methods are at the same *relative* heights in the hierarchy.

Although in this simple data set the two methods have given identical results, this is rarely the case in actual vegetation data. It has been found from trials conducted over many years that the Macnaughton-Smith Information Statistic provides results that are more ecologically sound than Association Analysis using chi-square.

Information Statistic (Lance & Williams 1968)

Although still a divisive monothetic method, the procedure is somewhat different from the previous ones. A more usual version of information statistic is used − a statistic which computes the total amount of information contained by a group of stands. When the group of stands is divided into two sub-groups, the information content of each sub-group is also computed; but *the sum of the information contents of the two sub-groups is always less than the information of the original group.* In other words, there is a loss of information whenever a group is divided into two sub-groups.

Let a group of stands (designated i) be divided into two sub-groups (g) and (h); then the fall in information content, ΔI, is defined as

$$\Delta I_{(gh,i)} = I_i - (I_g + I_h) \tag{6.2}$$

The information content of a group of stands is given by

$$I = sn \log_e n - \sum_{j=1}^{s} \{a_j \log_e a_j + (n - a_j)\log_e(n - a_j)\} \tag{6.3}$$

108

where s is the number of species contained in the group of n stands, and a_j is the number of stands containing the jth species.

Example 6.4

For all 10 stands of the Artificial Data, $s = 5$, $n = 10$, $a_I = 5$, $a_{II} = 6$, $a_{III} = 5$, $a_{IV} = 3$, $a_V = 1$; so

$$I = 50.\log_e 10 - \{(5.\log_e 5 + 5.\log_e 5) + (6.\log_e 6 + 4.\log_e 4)$$
$$+ (5.\log_e 5 + 5.\log_e 5) + (3.\log_e 3 + 7.\log_e 7)$$
$$+ (1.\log_e 1 + 9.\log_e 9)\} = 29.95$$

In using the idea of information loss on division of a group of stands, Equation 6.2, for a divisive monothetic classification, the strategy is to divide the stands into two sub-groups on the basis of the presence and absence of each species in turn, calculating the information loss each time. The species that governs the division of the stands into two sub-groups giving rise to the greatest loss of information is selected as the divisor species. In the next example, we shall work through two levels of division using the Artificial Data.

Example 6.5

Starting with all 10 stands, we first divide them into two sub-groups on the basis of presence and absence of Species I. We may designate the information content of the whole group simply as I, and the information contents of the two sub-groups as I_{I+} and I_{I-}; thus, in the notation of Equation 6.2, we have

$$\Delta I_I = I - (I_{I+} + I_{I-})$$

We already have $I = 29.95$ from Example 6.4; it remains to compute I_{I+} and I_{I-}, and the details of the Artificial Data set split on the presence and absence of Species I are as follows:

	I	II	III	IV	V		I	II	III	IV	V
1	1	0	1	0	0	3	0	0	1	0	1
2	1	1	0	0	0	5	0	1	0	1	0
4	1	1	0	1	0	7	0	0	0	1	0
6	1	1	1	0	0	8	0	0	1	0	0
9	1	1	0	0	0	10	0	1	1	0	0

(The two sub-groups are headed **I +** and **I −** respectively.)

For sub-group I + : $s = 4$, $n = 5$, $a_I = 5$, $a_{II} = 4$, $a_{III} = 2$, $a_{IV} = 1$; so

$$I_{I+} = 20.\log_e 5 - \{(5.\log_e 5 + 0.\log_e 0) + (4.\log_e 4 + 1.\log_e 1)$$
$$+ (2.\log_e 2 + 3.\log_e 3) + (1.\log_e 1 + 4.\log_e 4)\} = 8.37$$

For sub-group I − : $s = 4$, $n = 5$, $a_{II} = 2$, $a_{III} = 3$, $a_{IV} = 2$, $a_V = 1$; so

$$I_{I-} = 20.\log_e 5 - \{(2.\log_e 2 + 3.\log_e 3) + (3.\log_e 3 + 2.\log_e 2)$$
$$+ (2.\log_e 2 + 3.\log_e 3) + (1.\log_e 1 + 4.\log_e 4)\} = 12.60$$

Therefore

$$\Delta I_I = 29.95 - (8.37 + 12.60) = 8.98$$

Similar computations for the other species give: $\Delta I_{II} = 9.50$, $\Delta I_{III} = 11.49$, $\Delta I_{IV} = 9.51$, $\Delta I_V = 4.74$. Hence division on Species III gives the greatest fall in information content; and so, as in the previous methods, Species III is the first divisor species, with an information loss of 11.49. Both I_{III+} and I_{III-} equal 9.23.

Turning now to the first of the two sub-groups, we first divide on the basis of Species I giving the two sub-sub-groups:

III + , I +							III + , I −					
	I	II	III	IV	V		I	II	III	IV	V
1	1	0	1	0	0		3	0	0	1	0	1
6	1	1	1	0	0		8	0	0	1	0	0
							10	0	1	1	0	0

For group III + , I + : $s = 3$, $n = 2$, $a_I = 2$, $a_{II} = 1$, $a_{III} = 2$; so

$$I_{III+,I+} = 6.\log_e 2 - \{(2.\log_e 2 + 0.\log_e 0)$$
$$+ (1.\log_e 1 + 1.\log_e 1) + (2.\log_e 2 + 0.\log_e 0)\} = 1.39$$

For group III + , I − : $s = 3$, $n = 3$, $a_{II} = 1$, $a_{III} = 3$, $a_V = 1$; so

$$I_{III+,I-} = 9.\log_e 3 - \{(1.\log_e 1 + 2.\log_e 2)$$
$$+ (3.\log_e 3 + 0.\log_e 0) + (1.\log_e 1 + 2.\log_e 2)\} = 3.82$$

Hence

$$\Delta I_{III+,I} = 9.23 - (1.39 + 3.82) = 4.02$$

110

Exactly similar calculations follow for a split on Species II, and the result is the same $\Delta I_{III+,II} = 4.02$. Species III is present in all stands, and Species IV is absent from all stands, so those two species do not contribute anything here. For the split on Species V, we have only one stand, 3, containing this species; but inspection of Equation 6.3 shows that the information content of a single stand is zero. The information content of the four stands not containing Species V is 5.55; hence the information lost in the split is

$$\Delta I_{III+,V} = 9.23 - (0 + 5.55) = 3.68$$

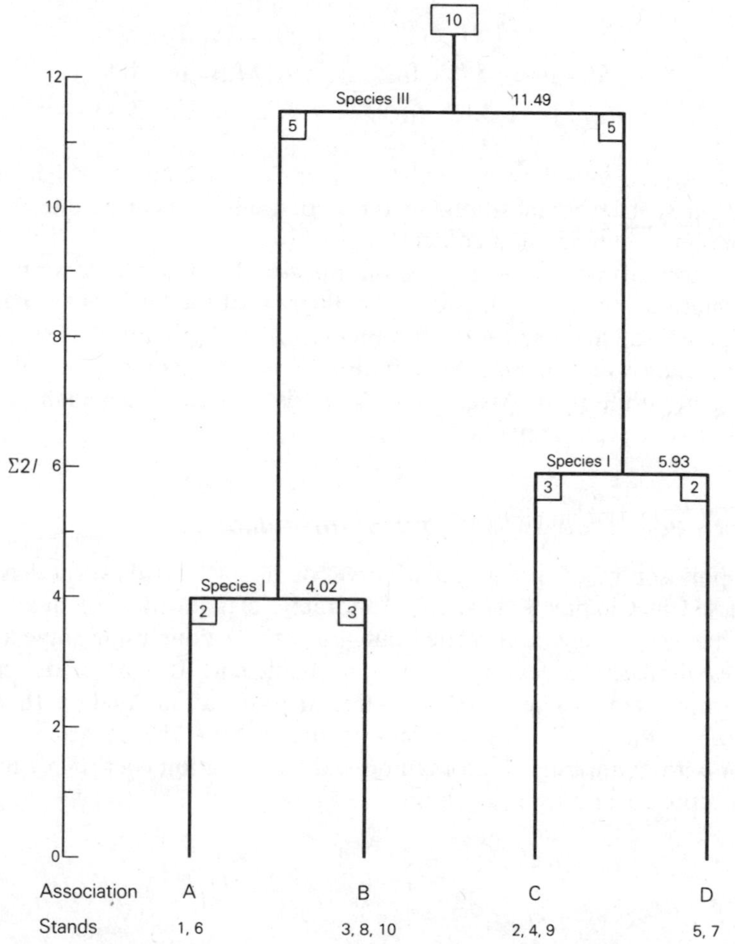

Figure 6.2 Normal Information Statistic (Lance & Williams 1968) analysis of the Artificial Data. Numbers of stands at each stage are shown in boxes, and the divisor species and $(\Sigma 2I)_{max}$-values are shown for each division.

111

Both splits on Species I and on Species II give an equal higher information loss than on Species V. Since there are no ecological features to guide us here, let us simply take the first of the two species, I, giving groups

III + , I + , i.e. stands 1, 6; and III + , I − , i.e. stands 3, 8, 10

bearing in mind that on an information loss criterion the split III + , II + , i.e. stands 6, 10, and III + , II − , i.e. stands 1, 3, 8, would do equally well.

The calculations for the second division of the III − stand group will be left as an exercise for you, and the results are merely quoted here.

$$I_{\mathrm{III}-,\mathrm{I}+} = 1.91, \ I_{\mathrm{III}-,\mathrm{I}-} = 1.39; \ \Delta I_{\mathrm{III}-,\mathrm{I}} = 5.93$$

$$I_{\mathrm{III}-,\mathrm{III}+} = 5.02, \ I_{\mathrm{III}-,\mathrm{II}-} = 0; \ \Delta I_{\mathrm{III}-,\mathrm{II}} = 4.21$$

$$I_{\mathrm{III}-,\mathrm{IV}+} = 3.82, \ I_{\mathrm{III}-,\mathrm{IV}-} = 0; \ \Delta I_{\mathrm{III}-,\mathrm{IV}} = 5.41$$

In the 'III − , IV − ' group there are just Stands 2 and 9 which have an identical species composition, so the two stands appear as one from the viewpoint of information content.

Division of the III − stands on Species I gives the largest fall of information content, 5.93, and so the diagram of the final result appears as in Figure 6.2. On comparison with Figure 6.1, a slight difference is evident; so the Lance & Williams Information Statistic method gives a slightly divergent result from Association Analysis and the Macnaughton-Smith Information Statistic methods.

Association Analysis using quantitative data

The philosophy of using quantitative data for classification has been discussed in Chapter 3. Divisive monothetic classification of quantitative data has very rarely been carried out, but can be done using some kind of species similarity index or correlation coefficient. Just as in the case of qualitative data, where χ^2 works better in Association Analysis than does $\sqrt{(\chi^2/n)}$ − the qualitative correlation coefficient − the covariance works better with quantitative data than does the correlation coefficient. For two species the covariance is given by

$$c = \sum_{i=1}^{n} x_i y_i - \left(\sum_{i=1}^{n} x_i \sum_{i=1}^{n} y_i \right) \Big/ n \tag{6.4}$$

where x_i and y_i are the abundance values of the first and second species, respectively, in the ith stand.

Example 6.6

Using the quantitative version of the Artificial Data, we first calculate a matrix of covariances from the species columns taken in all possible pairs. As with χ^2, we ignore the signs of the covariances

	I	II	III	IV	V	Σc
I	—					42.6
II	13.6	—				138.6
III	13.0	87.0	—			$155.5 = (\Sigma c)_{max}$
IV	14.8	33.8	49.0	—		99.0
V	1.2	4.2	6.5	1.4	—	13.3

As with qualitative Association Analysis, Species III has the highest index value, $(\Sigma c)_{max}$; so we have the same primary sub-groups as before. The first of these, III + , has the following structure:

	I	II	V
1	3	0	0
3	0	0	1
6	2	6	0
8	0	0	0
10	0	9	0

with covariance matrix:

	I	II	V	Σc
I	—			4.0
II	3.0	—		$6.0 = (\Sigma c)_{max}$
V	1.0	3.0	—	4.0

So the split is made on Species II, giving the groups:

III + , II + , i.e. Stands 6, 10, and III + , II − , i.e. Stands 1, 3, 8

The III − group comprises

	I	II	IV
2	1	10	0
4	1	2	2
5	0	7	3
7	0	0	9
9	5	8	0

113

and the covariance matrix is

	I	II	IV	Σc
I	—			31.8
II	14.2	—		64.8
IV	17.6	50.6	—	$68.2 = (\Sigma c)_{max}$

The divisor species is now IV, and so the two groups are

III − , IV + , i.e. Stands 4, 5, 7, and III − , IV − , i.e. Stands 2, 9

The whole analysis is shown in Figure 6.3.

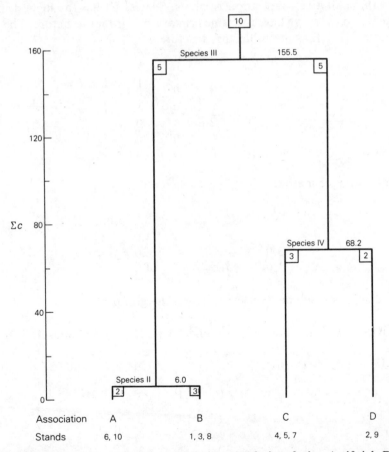

Figure 6.3 Normal quantitative Association Analysis of the Artificial Data. Numbers of stands at each stage are shown in boxes, and the divisor species and Σc-values are shown for each division.

114

Comparison of methods

All the four divisive monothetic classification methods presented agree on the first divisor species, but differences become apparent at the second level. Without any ecological information to guide us it is difficult to comment on the relative merits and disadvantages of the methods, but the following points may be made:

(a) Macnaughton-Smith's Information Statistic and χ^2 Association Analysis treat Species V as a singular one in that the only stand in which it occurs appears as an association of its own. Further, these methods are marginally better at limiting the number of species per association in the Artificial Data set. On the other hand, the division of III + stands is of relatively minor importance, occurring at a low level of $\Sigma\chi^2$ or $\Sigma 2I$ as the case may be.

(b) The quantitative Association Analysis is the only one to show the identical Stands 2 & 9 (from a species presence only point of view) as a separate association, although it must be realized that these two stands are not identical from a species abundance viewpoint.

(c) The Lance & Williams Information Statistic divides the III + stands at a relatively high level of ΔI, unlike the other two which indicate that the division is relatively unimportant.

A further comparison will be made later in the chapter (pp. 137–9); but you could now read about the applications of Macnaughton-Smith's Information Statistic to the two case studies in Chapter 9, pages 244 and 284.

Agglomerative polythetic methods

As previously remarked, any classification method involves a particular strategy combined with a specific measure of similarity. In divisive methods there is only one method of constructing hierarchical dichotomies: among the four divisive monothetic methods just presented, although the Lance & Williams Information Statistic procedure *appears* to differ from the others in the way by which a division is effected, the strategy of division is exactly the same.

In an overall agglomerative setting, however, not only may we use different similarity measures, but different fusion strategies exist also; hence, very many agglomerative polythetic methods can be formulated. It is not the intention to review these possibilities here, since agglomerative polythetic approaches are not as popular as they once were. Williams *et al.*

(1966) compared a number of such methods and found one particular combination of strategy and measure to be very markedly superior to the others. They called the method **Information Analysis**, and it appears in detail below.

Information Analysis (Williams, Lambert & Lance 1966)

In a way, this technique is the reverse of the Lance & Williams Information Statistic divisive monothetic method. Just as there is a loss of information when a group of stands is divided, so there is a gain of information when stands are amalgamated to form a group. In a divisive strategy we are attempting to make the split sub-groups as different as possible, hence we wish to *maximize* information loss. In fusion, we aim to join together those stands which are as alike as possible, and so we are looking for an amalgamation which *minimizes* the information gain.

As the name of the method implies, we use an information statistic as a measure, and it is the same one as used in the Lance & Williams Information Statistic method, Equation 6.3. The strategy of amalgamation is an example of **centroid fusion**, which implies that once two or more stands have been fused they are never again separated, but go up the agglomerative hierarchy as a group until they are subsequently fused with one or more other stands. It can be shown that a total of $n - 1$ fusions are necessary to build the complete hierarchy.

The only way to appreciate how the method works is to go through it, which we shall do in Example 6.7, using the Artificial Data. The presentation will seem much longer and more involved than the methods described hitherto because we must deal with the whole hierarchy, starting with the individual stands and ending with all ten amalgamated together.

Example 6.7

First, here is a demonstration of the use of Equation 6.3 by fusion of Stands 1 and 2. We have $s = 3$, $n = 2$, $a_I = 2$, $a_{II} = 1$, $a_{III} = 1$; so

$$I_{1,2} = 6 . \log_e 2 - \{(2 . \log_e 2 + 0 . \log_e 0) + 2(1 . \log_e 1 + 1 . \log_e 1)\} = 2.77$$

which is the information content of combined Stands 1 & 2. Since the information content of each stand separately is zero, the information gain by fusing Stands 1 and 2 is also 2.77.

To start the hierarchy of fusions, we need to calculate the information gain by amalgamating the stands in all possible pairs: there are $^{10}C_2 = 45$ of these fusions, and their information gains are listed in Table 6.3. Unsurprisingly, the lowest gain of information is given by fusing the two identical

Table 6.3 The first fusion set for the agglomerative Information Analysis of the Artificial Data.

$I_{1,2}\ = 2.77$	$I_{3,10} = 2.77$
$I_{1,3}\ = 2.77$	$I_{4,5}\ = 1.39 =$ 4th actual fusion
$I_{1,4}\ = 4.16$	$I_{4,6}\ = 2.77$
$I_{1,5}\ = 5.55$	$I_{4,7}\ = 2.77$
$I_{1,6}\ = 1.39 =$ 2nd actual fusion	$I_{4,8}\ = 5.55$
$I_{1,7}\ = 4.16$	$I_{4,9}\ = 1.39$
$I_{1,8}\ = 1.39$	$I_{4,10} = 4.16$
$I_{1,9}\ = 2.77$	$I_{5,6}\ = 4.16$
$I_{1,10} = 2.77$	$I_{5,7}\ = 1.39$
$I_{2,3}\ = 5.55$	$I_{5,8}\ = 4.16$
$I_{2,4}\ = 1.39$	$I_{5,9}\ = 2.77$
$I_{2,5}\ = 2.77$	$I_{5,10} = 2.77$
$I_{2,6}\ = 1.39$	$I_{6,7}\ = 5.55$
$I_{2,7}\ = 4.16$	$I_{6,8}\ = 2.77$
$I_{2,8}\ = 4.16$	$I_{6,9}\ = 1.39$
$I_{2,9}\ = 0\qquad =$ 1st actual fusion	$I_{6,10} = 1.39$
$I_{2,10} = 2.77$	$I_{7,8}\ = 2.77$
$I_{3,4}\ = 6.93$	$I_{7,9}\ = 4.16$
$I_{3,5}\ = 5.55$	$I_{7,10} = 4.16$
$I_{3,6}\ = 4.16$	$I_{8,9}\ = 4.16$
$I_{3,7}\ = 4.16$	$I_{8,10} = 1.39$
$I_{3,8}\ = 1.39 =$ 3rd actual fusion	$I_{9,10} = 2.77$
$I_{3,9}\ = 5.55$	

Stands 2 and 9, and is therefore zero. This is then the first actual fusion (Fig. 6.4a); it could have been anticipated from our existing knowledge of the data set, but identical stands (in the qualitative sense) are rather unusual in real field data.

It is important to realize that all the fusions in the sets in Tables 6.3 and 6.4 are *potential* fusions only. The *actual* fusion is the one giving the lowest gain of information, but subject to certain constraints.

Next, a second set of fusions is made. This is much shorter than the first, because we now only require information contents of fusions between Stands 2 & 9 taken together and each of the other eight stands. The modified data table is shown in the upper left-hand corner of Table 6.4, and the fusions based on the modified data table are shown to the right of it. At this stage, we require the fusion with lowest information content other than 2 & 9, and other than combinations involving 2 and 9 separately since they are now permanently fused together. To find the appropriate fusion we have to scan both the first and second fusion sets and select the minimum. In this small data set there are several possible candidate pairs at $I = 1.39$, and so we select the first one encountered – Stands 1 and 6 (Fig. 6.4b).

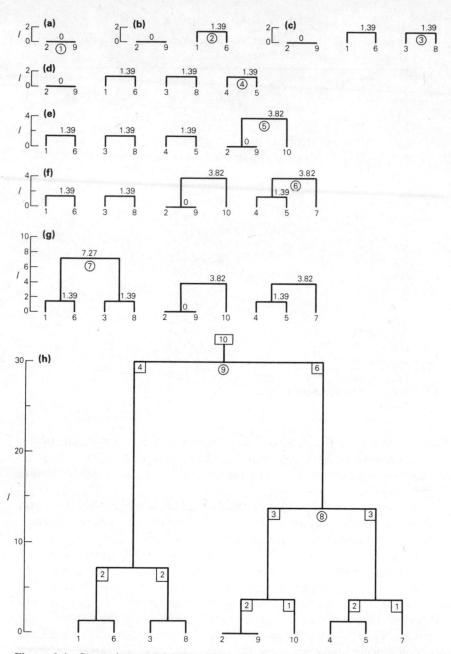

Figure 6.4 Stages in normal Information Analysis of the Artificial Data, starting with the individual stand fusions. The encircled numbers beneath the horizontal lines show the order of fusion, and the numbers above the lines show the information content, *I*, at that level. In (h), the numbers of stands at each stage are shown in boxes.

Next, a third fusion set is made by combining the latest amalgamation, 1 & 6, with 2 & 9 and the remaining single stands (3, 4, 5, 7, 8, 10), shown in the second section of Table 6.4. Now we work through all three fusion sets. Since there are still several appropriate amalgamations of single stands at $I = 1.39$, we merely ensure that none of the I's in the second or third fusion sets are less than 1.39: there aren't any! Accordingly, in the first fusion set the next value of 1.39 after Stands 1 & 6 and which does not involve Stands 1, 2, 6, or 9 as individuals, is selected; this is between Stands 3 & 8 (Fig. 6.4c).

The fourth fusion set involves amalgamating the latest fused pair, Stands 3 & 8, with 1 & 6, 2 & 9, 4, 5, 7, 10; the modified data table and relevant fusions are shown in the third section of Table 6.4. There is still an I-value of 1.39 in the first fusion set which is appropriate, and is lower than any I-value in subsequent fusion sets; we find it by moving on from $I_{3,8}$ in the first fusion set, and it is $I_{4,5}$. So our fourth actual fusion is between Stands 4 & 5 (Fig. 6.4d).

The fifth fusion set amalgamates Stands 4 & 5 with 1 & 6, 2 & 9, 3 & 8, 7, 10; relevant details are in the fourth section of Table 6.4. To make the actual fusion, we note that only Stands 7 and 10 now exist on their own, and that $I_{7,10} = 4.16$; but we must check in the other fusion sets to see if there is a lower value than this. There are $I_{(2,9),4}$ and $I_{(2,9),6} = 1.91$ in the second fusion set, but Stands 4 and 6 are no longer isolated. This brings us to a number of 3.82 values. The first two of these in the second fusion set are for Stands (2, 9) & 1, and Stands (2, 9) & 5. These fusions are no good because Stands 1 and 5 no longer exist as separate entities; but an amalgamation between (2, 9) and 10 is feasible, and for this a slight re-arrangement of our developing tree diagram is necessary (Fig. 6.4e).

Now, the sixth fusion set comprises joinings of (2, 9, 10) with (1, 6), (3, 8), (4, 5), 7; see the fifth section of Table 6.4. In finding the actual sixth fusion, we note that the first fusion set is now defunct because there are no two single stands remaining; further, the second fusion set is also finished because joint Stand (2, 9) no longer exists on its own. So we only have to examine the third to the sixth fusion sets. The lowest I-values are 3.82, but (1, 6) & 8, (1, 6) & 10, and (3, 8) & 10 will not do; but (4, 5) & 7 is suitable, and is therefore the sixth actual fusion (Fig. 6.4f).

The seventh fusion set involves (4, 5, 7) with (2, 9, 10), (1, 6), and (3, 8) in Table 6.4. In finding the seventh actual fusion, we see that the third fusion set is no longer available, but some items in the fourth set onwards are. Only the first amalgamation in the fourth set is available, and none of the fifth set is valid any longer. Further, only the first two of the sixth set, and all of the seventh set are available — six possibilities in all. The lowest of these is $I_{(1,6),(3,8)} = 7.27$ (Fig. 6.4g).

The eighth fusion set joins (1, 3, 6, 8) with (4, 5, 7) and (2, 9, 10), at the

Table 6.4 The second to eight fusion sets for the agglomerative Information Analysis of the Artificial Data.

Modified data table	Fusions

Second fusion set:

	I	II	III	IV	V
2 9	11	11	00	00	00
1	1	0	1	0	0
3	0	0	1	0	1
4	1	1	0	1	0
5	0	1	0	1	0
6	1	1	1	0	0
7	0	0	0	1	0
8	0	0	1	0	0
10	0	1	1	0	0

$I_{(2,9),1} = 3.82$
$I_{(2,9),3} = 7.64$
$I_{(2,9),4} = 1.91$
$I_{(2,9),5} = 3.82$
$I_{(2,9),6} = 1.91$
$I_{(2,9),7} = 5.73$
$I_{(2,9),8} = 5.73$
$I_{(2,9),10} = 3.82 = $ 5th a.f.*

Third fusion set

	I	II	III	IV	V
1 6	11	01	11	00	00
2 9	11	11	00	00	00
3	0	0	1	0	1
4	1	1	0	1	0
5	0	1	0	1	0
7	0	0	0	1	0
8	0	0	1	0	0
10	0	1	1	0	0

$I_{(1,6),(2,9} = 5.02$
$I_{(1,6),3} = 5.73$
$I_{(1,6),4} = 5.73$
$I_{(1,6),5} = 7.64$
$I_{(1,6),7} = 7.64$
$I_{(1,6),8} = 3.82$
$I_{(1,6),10} = 3.82$

Fourth fusion set

	I	II	III	IV	V
1 6	11	01	11	00	00
2 9	11	11	00	00	00
3 8	00	00	11	00	10
4	1	1	0	1	0
5	0	1	0	1	0
7	0	0	0	1	0
10	0	1	1	0	0

$I_{(1,6),(3,8)} = 7.27 = $ 7th a.f.*
$I_{(2,9),(3,8)} = 10.57$
$I_{(3,8),4} = 9.55$
$I_{(3,8),5} = 7.64$
$I_{(3,8),7} = 5.73$
$I_{(3,8),10} = 3.82$

Fifth fusion set

	I	II	III	IV	V
1 6	11	01	11	00	00
2 9	11	11	00	00	00
3 8	00	00	11	00	10
4 5	10	11	00	11	00
7	0	0	0	1	0
10	0	1	1	0	0

$I_{(1,6),(4,5)} = 10.04$
$I_{(2,9),(4,5)} = 5.02$
$I_{(3,8),(4,5)} = 12.82$
$I_{(4,5),7} = 3.82 = $ 6th a.f.
$I_{(4,5),10} = 5.73$

Sixth fusion set

	I	II	III	IV	V
2 9 10	110	111	001	000	000
1 6	11	01	11	00	00
3 8	00	00	11	00	10
4 5	10	11	00	11	00
7	0	0	0	1	0

$I_{(2,9,10),(1,6)} = 8.37$
$I_{(2,9,10),(3,8)} = 12.60$
$I_{(2,9,10),(4,5)} = 9.23$
$I_{(2,9,10),7} = 9.52$

Table 6.4 (*continued*)

Modified data table	Fusions

Seventh fusion set

	I	II	III	IV	V	
4 5 7	1 0 0	1 1 0	0 0 0	1 1 1	0 0 0	$I_{(4,5,7),(2,9,10)} = 13.73 = $ 8th a.f.[*]
2 9 10	1 1 0	1 1 1	0 0 1	0 0 0	0 0 0	$I_{(4,5,7),(1,6)}\quad = 13.46$
1 6	1 1	0 1	1 1	0 0	0 0	$I_{(4,5,7),(3,8)}\quad = 15.10$
3 8	0 0	0 0	1 1	0 0	1 0	

Eight fusion set

	I	II	III	IV	V	
1 3 6 8	1 0 1 0	0 0 1 0	1 1 1 1	0 0 0 0	0 1 0 0	$I_{(1,3,6,8),(4,5,7)} = 21.99$
4 5 7	1 0 0	1 1 0	0 0 0	1 1 1	0 0 0	$I_{(1,3,6,8),(2,9,10)} = 16.62$
2 9 10	1 1 0	1 1 1	0 0 1	0 0 0	0 0 0	

[*]a.f. = actual fusion

end of Table 6.4. The calculations in full are: $s = 5$, $n = 7$, $a_I = 3$, $a_{II} = 3$, $a_{III} = 4$, $a_{IV} = 3$, $a_V = 1$;

$$I_{(1,3,6,8),(4,5,7)} = 35.\log_e 7 - \{(3.\log_e 3 + 4.\log_e 4) + (3.\log_e 3 + 4.\log_e 4)$$
$$+ (4.\log_e 4 + 3.\log_e 3) + (3.\log_e 3 + 4.\log_e 4)$$
$$+ (1.\log_e 1 + 6.\log_e 6)\} = 21.99$$

and $s = 4$, $n = 7$, $a_I = 4$, $a_{II} = 4$, $a_{III} = 5$, $a_V = 1$

$$I_{(1,3,6,8),(2,9,10)} = 28.\log_e 7 - \{(4.\log_e 4 + 3.\log_e 3) + (4.\log_e 4 + 3.\log_e 3)$$
$$+ (5.\log_e 5 + 2.\log_e 2)$$
$$+ (1.\log_e 1 + 6.\log_e 6)\} = 16.62$$

Only three amalgamations are now possible: (4, 5, 7) with (2, 9, 10) from the seventh set, and the two from the eighth set. The first of these amalgamations, (4, 5, 7) with (2, 9, 10), gives the lowest *I*-value of 13.73 (Fig. 6.4h).

The final fusion can only be (1, 3, 6, 8) & (2, 4, 5, 7, 9, 10), but we would normally need to compute the *I*-value to place the final horizontal line in the correct position on the diagram. However, in this case, we know from Example 6.4 that for the whole data set $I = 29.95$. Figure 6.4h shows the completed analysis.

A stopping position is chosen by drawing a horizontal line at a selected level of *I*. Since we have used divisive monothetic methods on the Artificial Data to produce four groups, let us now also delimit four groups from the

121

Information Analysis, which we can do by putting a line at $I = 6$ (say). Although some similarity can be found between the final groups of this and each of the four divisive monothetic methods, there is quite a fundamental distinction at the top of the diagram in that the latter methods give equal groupings at the first dichotomy, whereas the present method does not. However, the stand groupings given by the initial division are very similar, only Stand 10 being anomalous.

Divisive polythetic methods

A divisive polythetic method would seem to be the ultimate in efficiency. A divisive method is superior to an agglomorative method because one is starting with the entire data set and successively fragmenting it, while in the latter we start with fragments and have to build up the whole. Further, a monothetic method utilizes the information from just one attribute (species) at each level of division, whereas a polythetic criterion uses the information carried by several attributes, possibly even all of them. However, we pay for such efficiency, either in the complexity of the algorithm necessitating a large and intricate computer program, or in the excessive length of time required to pursue such an analysis even with the speed of the modern digital computer.

In this part of the chapter, we shall examine two quite different procedures. The first was suggested over 20 years ago; it uses the information carried by all the species, and is conceptually and algorithmically simple. Its great drawback is the impractically enormous amount of computing time required. The second method is 13 years old, and can be rapidly executed on a modern computer. It is a two-stage process at each level of division, but only the first could be said to be truly of a 'polythetic' nature, and is in fact not a classificatory procedure at all – hence the quotation marks around 'polythetic'. The second stage is where the dichotomy is actually made, and here only a few species contribute to the partitioning of the stands. This divisive polythetic method, in a slightly modified form, is probably the most widely used vegetative classification method at the present time; its only disadvantage is the program size.

The method of Edwards & Cavalli-Sforza (1965)

Because this method is of only marginal practical utility, it hasn't even acquired a definitive name. The principle is quite simple. The n stands are divided into two sub-groups in all possible ways. For each pair of sub-groups, two within group sums of squares are calculated (one for each sub-group); and the pair of sub-groups having the lowest within groups sum of squares, expressed as the total of the two, is the dichotomy ultimately selected.

Minimizing the within sub-groups sum of squares is synonymous with maximizing the between sub-groups sum of squares; in other words, we are aiming to produce the two tightest clusters of points representing the sub-groups, with the greatest distance between them, in m-dimensional species space. This can only be achieved by trying every possible combination, and there are $2^{(n-1)} - 1$ of them. Thus even for the Artificial Data with just ten stands there are $2^9 - 1 = 511$ combinations.

Indicator Species Analysis (Hill, Bunce & Shaw 1975)

During the early 1970s, several divisive polythetic classificatory schemes were devised, all of which were based on information provided by the first axis of an ordination (Ch. 7). Although not strictly applicable, the term 'polythetic' does describe the fact that an ordination is based on the information provided by all species. The various methods put foward involve different strategies of dividing the stands after the initial ordering along the first ordination axis. One of these methods – **Indicator Species Analysis** – has eclipsed the others in its popularity, and we shall describe it in full detail under a number of sub-headings, based on the description provided by Hill *et al.* in their paper. Since the method relies on an ordination, a full understanding of Indicator Species Analysis can only be achieved by reading Chapter 7 first.

THE ORDINATION

The key feature of the first axis of an ordination is that it coincides with the direction of maximum spread of the data (see pp. 169 & 171). Thus it should reflect the maximum variation and possibly the most important vegetation gradient. Any ordination method will do, but the prototype Indicator Species Analysis employed Reciprocal Averaging Ordination as it was considered to be the most efficient method available at that time. This is probably true today, as far as ordering the stands on the first axis is concerned; and the popular program version of Indicator Species Analysis – TWINSPAN – employs Reciprocal Averaging to produce stand scores on the first axis.

THE CONCEPT OF INDICATOR SPECIES

Having given each stand a score on the first ordination axis we have, in effect, a linear sequence of points, each point representing a stand (Fig. 6.5). The stands are now divided into two groups at the centroid (centre of gravity) or, equivalently, the mean of the stand scores (line CA in

123

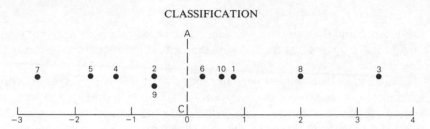

Figure 6.5 The positions of the ten stands of the Artificial Data on the first axis of a Reciprocal Averaging Ordination, with centroid (line CA).

Fig. 6.5). The stands on the left-hand side of the dividing line are referred to as the 'negative' group of the dichotomy, and those to the right as the 'positive' group; but it must be clearly understood that this does not refer to the absence or presence of any species, but simply to negtive and positive sides of the first axis if the centroid were considered to have the value zero.

If now the distribution of the species are examined in relation to their occurrence in the two groups, we would probably find that some species occurred exclusively in only one group, others occurred mainly in one group to a varying degree, while the remainder were largely indifferent to which side of the dividing line they occurred. Now if the first ordination axis really does reflect the major underlying floristic and hence environmental gradient, or a mix of the most important environmental gradients, then the positions of species tending to occur exclusively near one end or the other should be characteristic of environments at opposite ends of the gradient. These species are known as **indicator species**.

SELECTION OF A SPECIFIED NUMBER OF INDICATOR SPECIES

Most of the species in the entire set will have some indicator value, that is, some tendency to occur to one particular side of the dichotomy; but it is not essential, or even desirable, to use them all. Accordingly, a ranking of some measure of indicator value is required.

For the jth species, calculate its indicator value I_j, as

$$I_j = \frac{n_{j+}}{n_+} - \frac{n_{j-}}{n_-} \tag{6.5}$$

where n_+ is the total number of stands on the positive side of the dichotomy, and n_- similarly on the negative side (so that $n = n_+ + n_-$); n_{j+} is the number of stands on the positive side containing the jth species, and n_{j-} is the number of stands on the negative side containing the jth species. If a species occurs in every stand on the positive side, and in none at all on the negative side, $I_j = 1$; and conversely, if the species occurs in all stands on the negative side and in none at all on the positive side, $I_j = -1$. These are extreme situations, providing the best indicators. Hence the range of i_j is $-1 \leqslant I_j \leqslant 1$. Species for which $I_j = \pm 1$ are *perfect indicators*. Species occurring exclusively to one side, but not in all stands on that side will have $I_j < |\pm 1|$. So, for example, in 10 stands, 5 on either side, one species might occur in 2 stands only, both on the positive side, and would have $I_j = 0.4$; whereas another species might occur in 3 stands on the positive side and 1 stand on the negative side, and its I_j-value would also be 0.4 $(3/5 - 1/5)$. Thus, indicator values are partly determined by species frequency as well as by distribution.

In the prototype Indicator Species Analysis, five indicator species were used; these were the species with the five highest $|I_j|$-values, both positive and negative indicator values contributing equally.

STAND INDICATOR SCORES

Now each of the five indicator species are assigned the value $+1$ or -1 according to whether they have positive or negative I_j-values. Then each stand can be assigned an indicator value according to the numbers and signs of the indicator species the stand contains. For example, suppose that of the five indicator species four are positive and one is negative, and further suppose that a particular stand contains the negative indicator species and three of the positive indicator species, then the indicator value of that stand is $3 - 1 = 2$. The ranges of stand scores for the different possible distributions of five indicator species are shown in Table 6.5; in all cases there are six contingent scores, i.e one more than the number of indicator species.

DIVISION OF STANDS – THE INDICATOR THRESHOLD

In order to make a dichotomy, based now on indicator scores which in turn are based on indicator species content, we need to define an **indicator threshold**. This is defined as the maximum indicator score for a stand to be

Table 6.5 Range of stand indicator scores for the different possible distributions of five indicator species in Indicator Species Analysis.

Indicator species distribution		Possible indicator scores					
5 @ -1	0 @ $+1$	-5	-4	-3	-2	-1	0
4 @ -1	1 @ $+1$	-4	-3	-2	-1	0	1
3 @ -1	2 @ $+1$	-3	-2	-1	0	1	2
2 @ -1	3 @ $+1$	-2	-1	0	1	2	3
1 @ -1	4 @ $+1$	-1	0	1	2	3	4
0 @ -1	5 @ $+1$	0	1	2	3	4	5

included in the negative group; so that in the case of one negative and four positive indicator species, an indicator threshold of 1 gives stands of scores − 1, 0, 1 in the negative group, and scores of 2, 3, 4 in the positive group (Table 6.5). Apart from the highest indicator score of the appropriate set in Table 6.5, any of the other five scores could be used, and we select the one in which the stands on the negative and positive sides, as *defined by stand scores*, agree most closely with the distribution of the stands in the original dichotomy which was *based on the ordination*. The appropriate indicator score for this condition to be fulfilled is not necessarily at the mid-point, e.g. 1 for the situation of one negative indicator species in the analysis and four positive ones. Each of the five possible indicator thresholds must be tried in turn, and the best one, giving the minimum number of stands disagreeing in their placings according to the preliminary and final trial dichotomies, finally selected.

MISCLASSIFICATIONS

Even when the optimum indicator threshold has been selected there may still be one or more stands whose placings, on the basis of the division at the ordination axis' centroid and of division at the indicator threshold, disagree. Such stands are said to be *misclassified*. The assumption is that the original ordination score, being based on the whole species complement of the stand, is in general the 'correct' statement of the floristic composition of the stand. The indicator score, being based on only five species, is less reliable. Nevertheless, since our classification of stands is based on species content – the indicator species – we use the indicator score to define where the stand should be, while noting it as a misclassification if the stand's position is at variance with that suggested by the ordination.

BORDERLINE CASES AND THE 'ZONE OF INDIFFERENCE'

Many misclassifications, however, will merely be borderline, in that their original ordination scores conflict only marginally with their indicator scores. Now with a borderline case, the precise side of the division to which it is assigned is largely a matter of indifference; accordingly, a narrow **zone of indifference** is defined on the original ordination, within which all stands are classified according to their indicator score, whatever their ordination score.

The zone of indifference, in conjunction with the indicator threshold, is placed as follows. Let x_i' denote the score of the ith stand on the first axis of the ordination, and let

$$\bar{x} = \sum_{i=1}^{n} (a_{i.} x_i'/a_{..})$$

where $a_{i.}$ is the row total of the ith row of the data matrix (the total of all

126

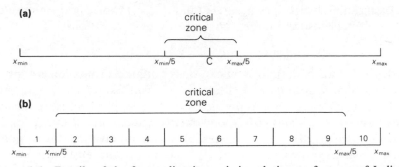

Figure 6.6 Details of the first ordination axis in relation to features of Indicator Species Analysis: (a) complete length of the ordination axis showing the critical zone around the centroid: (b) the same, but with magnification of the critical zone and diminution of the remainder of the axis. In (b), Segments 1 and 10 represent all parts of the axis outside the critical zone, and Segments 2 to 9 represent eight equal divisions of the critical zone. (After Hill *et al.* 1975, by permission of the *Journal of Ecology.*)

species scores in the ith stand), and

$$a_{..} = \sum_{i=1}^{n} a_{i.} \qquad \text{(the total of all the row totals)}$$

This weighted mean (point C in Figs 6.5 & 6.6a) is the natural zero point of the ordination (Hill 1974), and the ordination scores are therefore centred by subtracting the mean from each of them, i.e. $x_i = x_i' - \bar{x}$. Let x_{min} and x_{max} be the minimum and maximum stand scores, and a critical zone is demarcated to run from $x_{min}/5$ to $x_{max}/5$, as shown in Figure 6.6a. Then the critical zone is divided into eight equal segments (2–9 in Fig. 6.6b). The indicator species are selected, as already described, but with the proviso that species occurring in stands lying in the short length around the mean, defined by segments 4–7, are ignored as being too indeterminate to provide useful information on the indicator properties of the species.

A length of one-half of the critical zone length is used as the zone of indifference; this is therefore one-tenth of the total length of the first ordination axis from x_{min} to x_{max}. When the indicator scores of the stands have been calculated, the indicator threshold is determined by setting the zone of indifference to cover segments 4–7 of the critical zone, and computing the number of misclassifications which arise when each of the five possible values of the indicator threshold is used. The threshold is chosen so as to minimize this number. Finally, having selected the threshold, the five possible choices of the zone of indifference, namely segments 2–5, 3–6, 4–7, 5–8, 6–9, are compared for the number of misclassifications which they generate, and the best choice is again selected as that which yields the smallest number of misclassifications.

FINAL DICHOTOMY AND SUBSIDIARY CLASSES

Finally, at this stage, the division point on the original ordination is allowed to shift so that it accords as well as possible with the resulting indicator threshold. This is necessary, as the indicator threshold has only five possible values, none of which may correspond exactly to the centre of gravity of the ordination scores. There is no special virtue in dividing the original ordination at its centroid; it is merely convenient, and any other nearby point would do equally well. Accordingly, the ordination is finally divided at a point which agrees as well as possible with the chosen indicator threshold.

At the end of all this, we may have six categories of stand, whose relationships are shown in Figure 6.7.

(1) Negative group.
(2) Borderline negatives, assigned to the negative group because of their indicator scores.
(3) Misclassified negatives. Their ordination scores show that their affinities

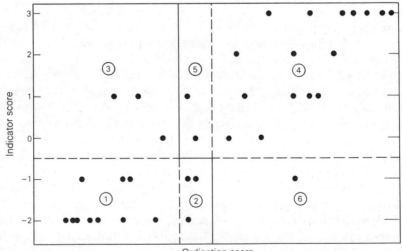

Figure 6.7 Scatter diagram showing the relationship between indicator score and ordination score for hypothetical data. The borderline cases are confined to regions 2 and 5 which is the zone of indifference; regions 3 and 6 contain misclassified stands. The indicator threshold in this case is − 1. For further explanation, see text. (After Hill *et al.* 1975, by permission of the *Journal of Ecology*.)

are unequivocally with the negative group, but their indicator scores place them in the positive group.

(4) Positive group.

(5) Borderline positives, assigned to the positive group because of their indicator scores.

(6) Misclassified positives. Their ordination scores show that their affinities are unequivocally with the positive group, but their indicator scores place them in the negative group.

It is to be hoped, and indeed usually is the case, that Groups 1 and 4 contain nearly all the stands, with extremely few in Groups 2, 3, 5, 6.

PREFERENTIAL SPECIES

Having made the dichotomy, the **preferential species** are tabulated. A species is deemed preferential to one side of the division if the relative frequency of its occurrence there is more than twice that of its occurrence in the other group. If the division is very unequal, then a species can be (weakly) preferential to the smaller group while still occurring in more stands of the larger group. For example, if a division is into 40 stands and 10, then a species which occurs in 12 out of the 40, and in 9 out of the 10, is deemed to be preferential to the smaller group as its relative frequency there is 0.9 as opposed to 0.3 in the larger group.

DOWNWEIGHTING OF RARER SPECIES

In initial trials of the method by Hill *et al.* (1975) it became apparent that Reciprocal Averaging Ordination was, for the purpose of this divisive classificatory strategy, too sensitive to the presence of a few aberrant stands containing rare species. In accordance with the view that the division should reflect major trends in the data, rather than detect outliers, a downweighting procedure for the rarer species was ultimately programmed into the ordination, as follows.

In the $n \times m$ data matrix, $\mathbf{X}$, where the rows represent stands and the columns species, the weight function is equivalent to multiplying the jth column by a factor w_j. If $a_{.j} = \sum_{i=1}^{n} x_{ij}$, the sum of the jth column, then the weighting applied to the jth species, w_j, is

$$\left. \begin{array}{l} w_j = 1 \text{ if } a_{.j} \geqslant m' \text{ (species not downweighted)} \\ w_j = (a_{.j}/m')^2 \text{ if } a_{.j} < m' \text{ (rare species downweighted)} \end{array} \right\} \quad (6.6)$$

where $m' = m/5$, and is taken to be the threshold of 'rarity'.

INDICATOR SPECIES ANALYSIS USING QUANTITATIVE DATA

With quantitative data there is, of course, no difficulty with the ordination;

129

the problem is to devise a classificatory method. The initial intuitive method of doing an ordination with quantitative data and then proceeding by the above method not only gives poor results, but is conceptually unsatisfying as the classification strategy uses only the species presences.

There is no 'natural' modification of the indicator species method to a quantitative strategy; any adjustment will have to be artificial, and for this purpose the concept of **pseudo-species** was developed by Hill *et al*. The idea is to introduce extra 'species', having the same name, to represent the abundance of a particular species; and the number of these pseudo-species required for any one species in a stand is proportional to the abundance of that species there. Naturally, there is a severe limit to the number of pseudo-species that can be defined, otherwise the data matrix would soon exceed the computer's memory, and a five-scale system involving five pseudo-species is commonly used. For example, if percentage cover were being used to assess the abundance of *Leontodon hispidus* (rough hawkbit) in stands on short turf calcareous grassland, the short logarithmic scale of Table 1.4 would do well, and we would have

---

L. hispidus 1 for cover up to 2%

L. hispidus 2 for cover 3–10%

L. hispidus 3 for cover 11–25%

L. hispidus 4 for cover 26–50%

L. hispidus 5 for cover 51–100%

These classes are non-exclusive, so that had only 2% cover been found in a stand it would be deemed to have only *L. hispidus* 1 in the analysis; whereas if in another stand the same species had 60% cover, all five pseudo-species would be deemed to occur in that stand.

There is one advantage and one disadvantage in this method of pseudo-species, but the latter can be made relatively trivial. The advantage is that the abundance value of species may now be indicators. If, for example, *L. hispidus* 3 was an indicator species, it would show that only a relatively substantial amount of *L. hispidus* has indicator value. On the other hand, the occurrence of two pseudo-species of the same species involves a redundancy; for example, *L. hispidus* 2 and *L. hispidus* 4 would show the former to be redundant. We thus have the additional rule that only one real species can be an indicator species, and this will be the highest pseudo-species number for that species.

THE HIERARCHY

All the above description relates to only one dichotomy; everything has to be repeated for each division made in the hierarchy. No way has yet been suggested for placing each division at an appropriate level of heterogeneity as for the monothetic methods; all the divisions at each level are placed along the same horizontal line. Neither is there an objective stopping rule, so the subjective method of specifying in advance the number of groups required has to be used.

Example 6.8

For the qualitative Artificial Data, we first calculate the weighted mean of the first ordination axis. The axis scores of the ten stands, calculated by Reciprocal Averaging Ordination, are

Stand 7: -2.68	Stand 6: 0.27
Stand 5: -1.73	Stand 10: 0.60
Stand 4: -1.28	Stand 1: 0.79
Stand 9: -0.58	Stand 8: 1.98
Stand 2: -0.58	Stand 3: 3.37

and the weighted mean (see p. 126 and Table 4.1) is given by:

$$\bar{x} = \frac{(1)(-2.68) + (2)(-1.73) + (3)(-1.28) + \cdots + (2)(3.37)}{20} = 0.05$$

The ordination dichotomy therefore separates Stands 2, 4, 5, 7, 9 with negative scores from Stands 1, 3, 6, 8, 10 with positive scores (Fig. 6.5).

Before calculating the indicator value of each species, we first require the critical zone from which is obtained the zone of indifference. We have $x_{min} = -2.68$, $x_{min}/5 = -0.54$; $x_{max} = 3.37$, $x_{max}/5 = 0.67$; and the two zones are shown in Figure 6.8. In particular, the zone of indifference

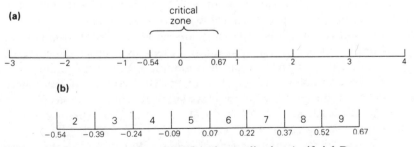

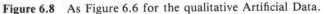

Figure 6.8 As Figure 6.6 for the qualitative Artificial Data.

131

stretches from -0.24 to 0.37; and Stand 6 alone, with a score of 0.27, lies within this range. From the data matrix (Table 4.1) we then calculate the species indicator scores, ignoring Stand 6. Hence in Equation 6.5, $n_+ = 4$ and $n_- = 5$; so the indicator value of each species is

$$I_I = \tfrac{1}{4} - \tfrac{3}{5} = -0.35$$

$$I_{II} = \tfrac{1}{4} - \tfrac{4}{5} = -0.55$$

$$I_{III} = \tfrac{4}{4} - \tfrac{0}{5} = 1.00$$

$$I_{IV} = \tfrac{0}{4} - \tfrac{3}{5} = -0.60$$

$$I_V = \tfrac{1}{4} - \tfrac{0}{5} = 0.25$$

Since there are only five species in total in this data set, we shall use them all as indicator species.

Next, bearing in mind that each species with a negative indicator value scores -1 and each species with a positive indicator value scores $+1$, the indicator score for each stand is calculated:

Stand 1: $-1 + 1$ $= 0$ Stand 6: $-1 - 1 + 1 = -1$

Stand 2: $-1 - 1$ $= -2$ Stand 7: -1 $= -1$

Stand 3: $+1 + 1$ $= 2$ Stand 8: $+1$ $= 1$

Stand 4: $-1 - 1 - 1 = -3$ Stand 9: $-1 - 1$ $= -2$

Stand 5: $-1 - 1$ $= -2$ Stand 10: $-1 + 1$ $= 0$

and, arranged in ascending order, the stand indicator scores are:

Stand:	4	2	5	9	6	7	1	10	8	3
Score:	-3	-2	-2	-2	-1	-1	0	0	1	2

Now, keeping the zone of indifference from -0.24 to 0.37, and setting the indicator threshold successively from -3 to 1, we find the number of misclassifications in each case. In Table 6.6, Stand 6 is asterisked to show that it does not count if it is apparently misclassified, since it occurs in the zone of indifference. An italicized number denotes a misclassified stand. Inspection of Table 6.6 shows that the optimum indicator threshold is -1.

Finally, we examine the effect of changing the position of the zone of indifference on the number of misclassifications while still keeping the indicator threshold at -1 (Table 6.7). Any of the three zones of indifference, 4–7, 5–8 & 6–9, are equally good.

We can now draw the same type of diagram as Figure 6.7 for the current

132

Table 6.6 The different dichotomies for each possible indicator threshold for the first division of the Indicator Species Analysis of the qualitative Artificial Data. Italicized numbers denotes misclassified stands, while an asterisk denotes occurrence in the zone of indifference.

Indicator threshold	Stands	Number of misclassifications
−3	4 *2,5,9,6*,7,1,10,8,3*	4
−2	4,2,5,9 6*,7,1,10,8,3	1
−1	4,2,5,9,6*,7 1,10,8,3	0 (6 is irrelevant)
0	4,2,5,9,6*,7,*1,10* 8,3	2 (6 is irrelevant)
1	4,2,5,9,6*,7,*1,10,8* 3	3 (6 is irrelevant)

situation (Fig. 6.9), and list the stands in their groups:

Negative group: 2, 4, 5, 6, 7, 9

Borderline negative: 6

Positive group: 1, 3, 8, 10

There are no misclassifications, either positive or negative, or borderline positives in the Artificial Data. You will notice that Stand 6 appears both in the negative group as well as the borderline negative group. This is necessary, because for a hierarchical classification all stands must finally appear in either the positive or negative group in order to be available for

Table 6.7 The effect of changing the zone of indifference, with an indicator threshold of −1, for the first division of the Indicator Species Analysis of the qualitative Artificial Data. Italicized numbers denote misclassified stands, while an asterisk denotes occurrence in the zone of indifference.

Zone of indifference	Stands	Number of misclassifications
2–5	4,2,5,9,6,7 1,10,8,3	1 (6 is not now in the zone of indifference)
3–6	4,2,5,9,*6*,7 1,10,8,3	1 (as above)
4–7	4,2,5,9,6*,7 1,10,8,3	0 (6 is now in the zone of indifference)
5–8	4,2,5,9,6*,7 1,10,8,3	0 (6 is in the zone of indifference)
6–9	4,2,5,9,6*,7 1,10*,8,3	0 (10 is now also in the zone of indifference)

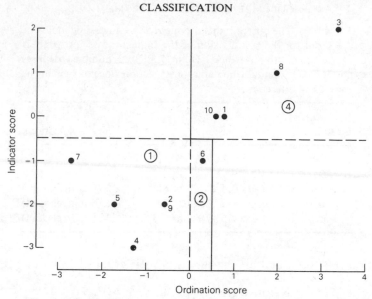

Figure 6.9 As Figure 6.7 for the qualitative Artificial Data.

the next division. The listing of a stand in a borderline or misclassification category should be regarded merely as supplementary information.

Negative preferential species are: Species I (occurs 4 times in the negative group: 1 time in the positive group), Species II (5 : 1) and Species IV (3 : 0); and positive preferential species are Species III (1 : 4) and Species V (0 : 1); so all the five species are preferential.

Example 6.9

The quantitative Artificial Data are now worked through. The scores of the ten stands on the first axis of the quantitative Reciprocal Averaging Ordination are:

Stand 3: −2.40	Stand 9: 0.29
Stand 8: −2.33	Stand 2: 0.50
Stand 1: −1.78	Stand 5: 1.73
Stand 10: −0.59	Stand 4: 1.99
Stand 6: 0.08	Stand 7: 4.47

and the mean is

$$\bar{x} = \frac{(11)(-2.40) + (9)(-2.33) + \cdots + (9)(4.77)}{12 + 11 + \cdots + 15} = 0.001$$

134

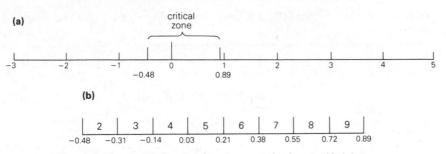

Figure 6.10 As Figure 6.6 for the quantitative Artificial Data.

or 0 to two decimal places. Further, $x_{min} = -2.40$, $x_{min}/5 = -0.48$; $x_{max} = 4.47$, $x_{max}/5 = 0.89$; so the critical and indifference zones are as shown in Figure 6.10. The zone of indifference stretches from -0.14 to 0.55, and Stands 6, 9, 2 lie within it. Now the stand indicator scores are calculated, ignoring information by the three stands in the zone of indifference. It is more of an involved job, however, than in the qualitative case, because we must now include all the pseudo-species. First, we must decide on the number of pseudo-species required and then determine their

Table 6.8 The quantitative Artificial Data matrix, showing the occurrence of pseudo-species.

	Pseudo-species																		
Stand	I 1	I 2	I 3	II 1	II 2	II 3	II 4	II 5	III 1	III 2	III 3	III 4	III 5	IV 1	IV 2	IV 3	IV 4	IV 5	V 1
Negative group																			
3	0	0	0	0	0	0	0	0	1	1	1	1	1	0	0	0	0	0	1
8	0	0	0	0	0	0	0	0	1	1	1	1	1	0	0	0	0	0	0
1	1	1	0	0	0	0	0	0	1	1	1	1	1	0	0	0	0	0	0
10	0	0	0	1	1	1	1	1	1	1	1	0	0	0	0	0	0	0	0
Positive group																			
5	0	0	0	1	1	1	1	0	0	0	0	0	0	1	1	0	0	0	0
4	1	0	0	1	0	0	0	0	0	0	0	0	0	1	0	0	0	0	0
7	0	0	0	0	0	0	0	0	0	0	0	0	0	1	1	1	1	1	0
Indifferent group																			
2	1	0	0	1	1	1	1	1	0	0	0	0	0	0	0	0	0	0	0
6	1	0	0	1	1	1	0	0	1	0	0	0	0	0	0	0	0	0	0
9	1	1	1	1	1	1	1	0	0	0	0	0	0	0	0	0	0	0	0

Table 6.9 Pseudo-species indicator scores evaluated from the upper two blocks of Table 6.8.

Pseudo-species	I 1	I 2	II 1	II 2	II 3	II 4
Indicator value	0.08	−0.25	0.42	0.08	0.08	0.08
Pseudo-species	II 5	III 1	III 2	III 3	III 4	III 5
Indicator value	−0.25	−1.00	−1.00	−1.00	−0.75	−0.75
Pseudo-species	IV 1	IV 2	IV 3	IV 4	IV 5	V 1
Indicator value	1.00	0.67	0.33	0.33	0.33	−0.25

'cut' levels. The abundance scale used in the data matrix (Table 4.1) runs from 1 to 10; thus, if we use five pseudo-species for equal intervals on the scale we may define the pseudo-species cut levels as follows:

Pseudo-species no.	Abundance scale range	Suggested cut level
1	1, 2	0.1
2	3, 4	2.1
3	5, 6	4.1
4	7, 8	6.1
5	9, 10	8.1

Using the data in Table 4.1, and arranging the stands into two groups indicated by the ordination while ignoring the stands in the zone of indifference, we have the situation shown in Table 6.8. From this table, the pseudo-species indicator scores are computed, and are displayed in Table 6.9. Utilizing the rule that only one pseudo-species (the highest) of a given real species can be used as an indicator species, we have the following

Table 6.10 The different dichotomies for each possible indicator threshold for the first division of the Indicator Species Analysis of the quantitative Artificial Data. Italicized numbers denote misclassified stands, while an asterisk denotes occurrence in the zone of indifference.

Indicator threshold	Stands	Number of misclassifications
−2	1,3 *8*,9*,*10*,2*,6*,7,4,5	2
−1	1,3,8 9*,*10*,2*,6*,7,4,5	1
0	1,3,8,9*,10 2*,6*,7,4,5	0 (9 is irrelevant)
1	1,3,8,*9*,10,2*,*6*,7 4,5	1 (2,6 and 9 are irrelevant)

indicator species in descending absolute order:

III 3	IV 1	II 1	I 2	V 1
−1.00	1.00	0.42	−0.25	−0.25

This indicates that while small presences of Species II, IV, and V influence the dichotomy (particularly IV), Species I needs to be present in somewhat higher amount, and Species III must occur to quite a moderate degree before they influence the division.

Now calculate the stand indicator scores, using the data layout in Table 6.8; in ascending order we get

Stand:	1	3	8	9	10	2	6	7	4	5
Score:	−2	−2	−1	0	0	1	1	1	2	2

Now we determine the optimum indicator threshold, while holding the zone of indifference to the range −0.14 to 0.55 (Table 6.10). We see that the optimum indicator threshold is 0. Since there are no misclassifications, there is no point in trying different positions of the zone of indifference.

The diagram of the same type as Figure 6.7 now appears as in Figure 6.11, and the list of stands appears thus:

Negative group: 1, 3, 8, 9, 10

Borderline negative: 9

Positive group: 2, 4, 5, 6, 7

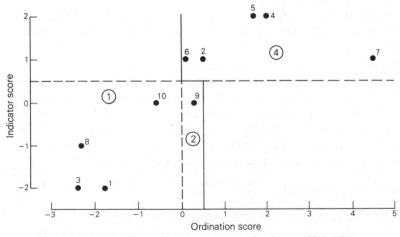

Figure 6.11 As Figure 6.7 for the quantitative Artificial Data.

Finally, we use Table 6.8 again to evaluate the preferential species. Negative preferentials are Species I 2 (2:0), Species I 3 (1:0), Species III 1 (4:1), Species III 2 (4:0), Species III 3 (4:0), Species III 4 (3:0), Species III 5 (3:0) and Species V 1 (1:0). Positive preferentials are Species IV 1 (0:3), Species IV 2 (0:2), Species IV 3 (0:1), Species IV 4 (0:1) and Species IV 5 (0:1). Thus, Species II does not figure as a preferential species, but the other four species do.

Comparison of the methods by the examples results

A full statement of the results of applying several classification methods to the Artificial Data, from the viewpoint of the final stand groupings, are presented in Table 6.11. The results indicate very well the range of method to method variation encountered in practice. Thus, while there are only minor disagreements between methods at the first level of divisions, discrepancies become more pronounced as one moves down the hierarchy. However, even at the second level there are a number of constant patterns; for example, Stands 1 and 2 are never in the same group, and Stands 5 and 7 are always together.

The Artificial Data are quite arbitrary, in the sense that they were put together without any underlying structure merely to illustrate numerically the workings of methods. So it is not surprising that different classification methods, whose strategies and criteria differ in the 'weighting' they give to the various features of the data, give a diversity of results. Real vegetation data tend to have rather more of a 'structure'; even if one does not believe in the community idea with its connotations of fairly rigid constitutions, nevertheless there is still some degree of constancy of species assemblages which the Artificial Data lack.

Perhaps the most surprising feature of the Artificial Data analyses is the similarity of results between methods based on monothetic and polythetic criteria. Only the place of insertion of Stands 6 and 9 differ between methods based on these two different criteria at the first level, and Indicator Species Analysis suggests that Stand 6 is borderline anyway. Admittedly, there are more discrepancies at the second level, but no more than exist between the two monothetic methods of Association Analysis (or Macnaughton-Smith's Information Statistic) and Lance & Williams' Information Statistic. Very similar remarks and comparisons can be made between analyses based on qualitative and quantitative data.

Inverse classifications

Normal (stand) and inverse (species) classifications

All that has been described for classifying stands into relatively homogen-

138

eous groups can also be applied, without change, to species. The same methods are applied to the same concepts, it is just that everything is reversed. Thus, whenever a row operation is carried out on a data matrix as a step in a stand classification method, the same now becomes a column operation as a step in the corresponding species classification, and vice versa; whenever we have used the word 'stand' in the description of a normal classification, we now substitute the word 'species' for an inverse classification, and vice versa. Every normal (or stand) classification method has its corresponding inverse (or species) classification, and we need only

Table 6.11 Comparison of several classification methods by the results (stands in groups) of their application to the Artificial Data. Group designations are arbitrary.

First level	Groups	
	A	B
Association Analysis		
Information Statistic (Macnaughton-Smith)	1, 3, 6, 8, 10	2, 4, 5, 7, 9
Information Statistic (Lance & Williams)	1, 3, 6, 8, 10	2, 4, 5, 7, 9
Information Analysis	1, 3, 6, 8	2, 4, 5, 7, 9, 10
Indicator Species Analysis	1, 3, 8, 10	2, 4, 5, 6, 7, 9
Quantitative Association Analysis	1, 3, 6, 8, 10	2, 4, 5, 7, 9
Quantitative Indicator Species Analysis	1, 3, 8, 9, 10	2, 4, 5, 6, 7

Second level	Groups			
	A	B	C	D
Association Analysis				
Information Statistic (Macnaughton-Smith)	1, 6, 8, 10	3	2, 4, 9	5, 7
Information Statistic (Lance & Williams)	1, 6	3, 8, 10	2, 4, 9	5, 7
Information Analysis	1, 6	3, 8	2, 9, 10	4, 5, 7
Indicator Species Analysis	1, 10	3, 8	2, 6, 9	4, 5, 7
Quantitative Association Analysis	6, 10	1, 3, 8	2, 9	4, 5, 7
Quantitative Indicator Species Analysis	9, 10	1, 3, 8	2, 6	4, 5, 7

describe one inverse method to illustrate the comparison between a particular type of stand and species classification. The terms 'normal' and 'inverse' were applied to Association Analysis when the method was proposed, and the more straightforward nomenclature of stand and species classifications were then scarcely used. However, the former terms were applied much less to later classification methods, and not at all to ordinations; hence, the terms 'normal' and 'inverse' are becoming redundant.

Association Analysis (Williams & Lambert 1961)

Starting with the data matrix, χ^2 is calculated for all possible stand pairs by the method shown on page 82; in terms of the data matrix we calculate a value of χ^2 for all possible pairs of rows. The total number of χ^2-values is $^nC_2$ (see p. 98), and two examples of the calculation have already been given (p. 83). The $n(n-1)/2$ chi-square values are arranged in an $n \times n$ matrix, and a sum chi-square is obtained for each stand. The stand with the highest $\Sigma\chi^2$ is designated the divisor stand, and the group of m species is divided into two sub-groups on the basis of whether or not a species does or does not occur in the divisor stand. Example 6.11 shows the working of the method in relation to the Artificial Data.

Example 6.10

As there are 10 stands, there will be $^{10}C_2 = (10)(9)/2 = 45$ χ^2-values to compute, which are then written out in the form of a 10×10 symmetric matrix:

	1	2	3	4	5	6	7	8	9	10	$\Sigma\chi^2$
1	—										15.90
2	0.14	—									13.74
3	2.22	2.22	—								16.87
4	5.00	2.22	5.00	—							$19.65 = (\Sigma\chi^2)_{max}$
5	2.22	0.14	2.22	2.22	—						12.01
6	2.22	2.22	0.14	0.14	2.22	—					14.09
7	0.83	0.83	0.83	0.83	1.88	1.88	—				9.05
8	1.88	0.83	1.88	1.88	0.83	0.83	0.31	—			11.15
9	1.25	5.00	2.22	0.14	0.14	2.22	0.83	0.83	—		12.77
10	0.14	0.14	0.14	2.22	0.14	2.22	0.83	1.88	0.14	—	7.85

Stand 4 has the highest sum chi-square and so it is the divisor stand; consequently, the species are divided into two groups: '4 + ', those species that occur in Stand 4, namely Species I, II, IV; and '4 − ', those species not occurring in Stand 4, i.e. Species III, V.

We now repeat the process on the first subgroup. First, the data for the species of Group 4 + are displayed:

	I	II	IV
1	1	0	0
2	1	1	0
5	0	1	1
6	1	1	0
7	0*	0	1
9	1	1	0
10	0	1	0

for which the χ^2 matrix is

	1	2	5	6	7	9	10	$\Sigma\chi^2$
1	—							6.75
2	0.75	—						11.25
5	3.00	0.75	—					6.75
6	0.75	3.00	0.75	—				11.25
7	0.75	3.00	0.75	3.00	—			11.25
9	0.75	3.00	0.75	3.00	3.00	—		11.25
10	0.75	0.75	0.75	0.75	0.75	0.75	—	4.50

Because we are working with such small numbers of species, there are four stands having the highest sum chi-square. However, selection of any of these stands, namely 2, 6, 7, 9, lead to the same result: that Species I and II form one group and Species IV forms a group on its own. This is fairly evident from inspection of the data matrix. The diagram of the whole analysis is shown in Figure 6.12. It is customary to show a species classification 'on its side' in order to facilitate the listing of species in the groups.

141

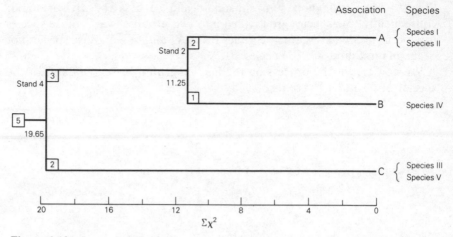

Figure 6.12 Inverse Association Analysis of the Artificial Data. Numbers of species at each stage are shown in boxes, and the divisor stands, together with $(\Sigma\chi^2)_{max}$-values, are shown for each division.

Information Analysis

By way of a comparison, the diagrammatic result of an Inverse Information Analysis is shown in Figure 6.13; the calculations are left as an exercise to the reader as they are not laborious. It is clear that the results of the two

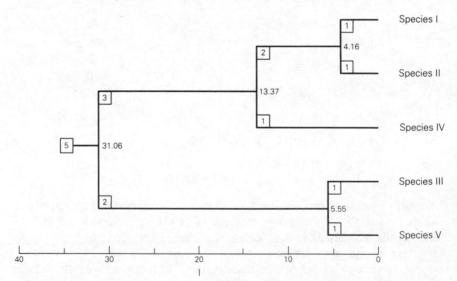

Figure 6.13 Inverse Information Analysis of the Artificial Data. Numbers of species at each stage are shown in boxes, and the divisor stands, together with values, are shown for each fusion.

Table 6.12 Two stages in Nodal Analysis of the Artificial Data: (a) primary two-way table of species (rows) and stands (columns) from inverse and normal Association Analyses; (b) the same, but rearranged to emphasize a diagonal of species-in-stand presences from top left to bottom right.

(a)

	3	1	6	8	10	2	4	9	5	7
I	0	1	1	0	0	1	1	1	0	0
II	0	0	1	0	1	1	1	1	1	0
IV	0	0	0	0	0	0	1	0	1	1
III	1	1	1	1	1	0	0	0	0	0
V	1	0	0	0	0	0	0	0	0	0

(b)

	2	4	9	5	7	1	6	8	10	3
I	1	1	1	0	0	1	1	0	0	0
II	1	1	1	1	0	0	1	0	1	0
IV	0	1	0	1	1	0	0	0	0	0
III	0	0	0	0	0	1	1	1	1	1
V	0	0	0	0	0	0	0	0	0	1

inverse classifications are essentially the same: indeed, there is very little room for variation as only five species are being classified.

Nodal Analysis (Lambert & Williams 1962)

We end this chapter on classification with a few notes on the technique of **Nodal Analysis**, but without going into detail. Nodal analysis is a method for combining a normal and inverse classification of the same type performed on a data matrix to produce units called **noda**. Noda are conceived to be species-in-habitat coincidences, and an approach to a realization of this concept may be obtained by combining the groups obtained in the normal and inverse analyses in a two-way table.

Table 6.12a shows a primary tabulation of the results of normal and inverse Association Analysis of the Artificial Data, in the order that the groups appear from the respective analyses. The boxes in the table are the noda, and the analysis attempts to put some degree of structure into the

Table 6.13 Final Nodal Analysis table based on Information Analysis.

	2	9	10	4	5	7	1	6	3	8
I	1	1	0	1	0	0	1	1	0	0
II	1	1	1	1	1	0	0	1	0	0
IV	0	0	0	1	1	1	0	0	0	0
III	0	0	1	0	0	0	1	1	1	1
V	0	0	0	0	0	0	0	0	1	0

table by re-arrangement of either the stand associations, or the species associations, or both. If species and stand groups were entirely coincident, that is, assuming no floristic variation from strict species-in-habitat coincidences and a perfect classification method, then some boxes in the table would be entirely filled with presences (1s) and the remaining boxes completely filled with absences (0s). For nearly all vegetation data there will be far more absences than presences, and the best structure one can impose on the two-way table is to have the noda containing mostly presences running from top left to bottom right in a broad band. Naturally, there is always variability around any strict species-in-habitat coincidences, and no classification can be perfect; so in practice the aim is to manoeuvre to the best possible table, always keeping the diagonal aim in view.

Table 6.12b shows the re-arranged tabulation where the objective has been broadly achieved for the Artificial Data. Stands 2, 4, 9 containing Species I and II is a very good nodum, as are Stands 5 and 7 comprising Species IV and Stand 3 with Species III and V. Stands 1, 6, 8, 10 don't appear to fit in the scheme quite as well. Stands and species within groups may also be changed round to improve the appearance, but this does not change the fundamental arrangement. Thus Stands 4 and 9 could be interchanged in Table 6.12b so as to bring the occurrence of Species IV in the former stand closer to the occurrences of this species in Stands 5 and 7.

Table 6.13 shows the Nodal Analysis for the Artificial Data based on Information Analysis. The result is not quite as good as for Association Analysis, and so Nodal Analyses could well feature in comparisons of classificatory methods for the purpose of finding the best classification of a particular set of data.

7

Ordination

Although part of the results set of a vegetation classification may be graphical – the hierarchical diagram – the main product is a list of stands (or species) occurring in the different groups, or associations. Ordination results, on the other hand, are almost wholly graphical; much of the computer print-out consists of lists of co-ordinate values from which the graphs are finally plotted.

Just as each set of vegetation (species-in-stands) data gives rise to two classifications – a stand (or normal) classification, and a species (or inverse) classification – so there are corresponding ordinations; but the terms 'normal' and 'inverse' have never been used in this context. Two, equally ambiguous, terms are employed to categorize ordinations: R-techniques correspond to stand ordinations, and Q-techniques designate species ordinations.

Ordination results show the stands or species plotted against two or sometimes three axes, each axis corresponding to a dimension in space. An ordination method aims to depict the gradient of greatest variation along the 'first' axis, the second largest gradient of variation along the 'second' axis, and so on. This is the *aim* of an ordination method; but the success of the different methods in achieving this aim is very variable, and to some extent depends on the structure of the set of data in hand. Axes 'higher' than the third may also be considered and graphed, but they are less useful since not only do they contain progressively less information, but there is very often a distorting effect of lower axes (say the first) on higher axes (e.g. the second); in other words, the positions of stands in relation to the second and subsequent axes is partly determined by the mathematical properties of the ordination method, and partly determined by the stands' 'true' places in relation to the gradient represented by the axis.

The use of the word 'gradient' introduces the central theme or idea in ordination. Axes on ordination diagrams may have apparently arbitrary scales, but nevertheless represent important trends. Whether these trends correspond to some ecological realities, or whether they arise as a result of the mathematics of the ordination method, varies with the actual method applied to a specific set of data. Usually the axes are a blend of both these features, and a good ordination method should emphasize ecological features and minimize trends due to mathematical artefacts.

An ordination of vegetation data will obviously reveal floristic trends or gradients to a greater or lesser degree. However, it is an axiom in ecology that there are correlations between species occurrences (and, to a lesser extent, abundances) and levels of environmental factors. Very often the occurrence of a species is governed by the environment, but not necessarily; sometimes an environmental factor in a stand is governed by the species present. An example of the former, involving soil factors, is that many species occurrences are governed by the calcium content of the soil; in particular, *Calluna vulgaris* is excluded from soils of high available calcium. However, if a seed of *Calluna vulgaris* happens to land and germinate on a small area of low calcium content in an otherwise highly calcareous soil, then growth of the established seedling results in its litter reducing the pH of the soil beneath; and the calcium level of this soil, already rather low, will become further depleted through leaching. Here, calcium and pH levels of the soil are reduced by the presence of *Calluna vulgaris*. Whichever direction the cause and effect takes, the fact remains that gradients in the floristic composition of stands will reflect gradients in environmental factors; so although the ordination is typically derived from vegetation data, and the axes represent gradients in the floristic composition of the stands involved, it is very likely that the axes reflect gradients in environmental factors also.

A classification of ordinations

A useful, although perhaps not universally acceptable, terminology is to designate ordinations of all kinds as **gradient analyses.** We then define a **direct gradient analysis** as an ordination based directly on environmental factors; the ordination axes are either individual environmental factors, or are combinations of several environmental factors which have been obtained by definite mathematical procedures. By contrast, an **indirect gradient analysis** is based on vegetation data; the ordination axes define gradients in the vegetation but, as already discussed, these gradients should reflect environmental gradients. Hence, the terms 'direct' and 'indirect' refer to ordination activities in relation to environmental factors.

In the following sub-sections, only certain concepts and generalized results will be given in order to build up a picture of the relationships between different ordination methods, and of the rationale behind them, in a sequence from the simplest to the most complicated. Detailed descriptions of the methods will follow later.

Direct Gradient Analysis – one factor

Direct Gradient Analysis involving just one environmental factor is the very simplest ordination technique. It is so elementary that a stand ordination is trivial, in that it is simply a linear sequence of points from left to right, reflecting the increasing value of the environmental factor represented by the axis (Fig. 7.1a); but a species Direct Gradient Analysis is particularly informative, and moreover forms a very important conceptual background for ordination methods in particular, and plant ecology in general.

One way of presenting the result of a Direct Gradient Analysis is shown in Figure 7.1b. Essentially, the smooth curves are obtained from stand data – each stand consisting of species abundance values and a measured level of the environmental factor concerned. The smooth curves are an idealization in relation to actual data; if done objectively by statistical methods, the curve fitting is by far the most difficult part of the analysis. More frequently, eye-fitted curves are drawn among the data points or, if the plotted points are relatively few and regular, adjacent ones may be joined by straight lines. An example of this is shown in Figure 9.19.

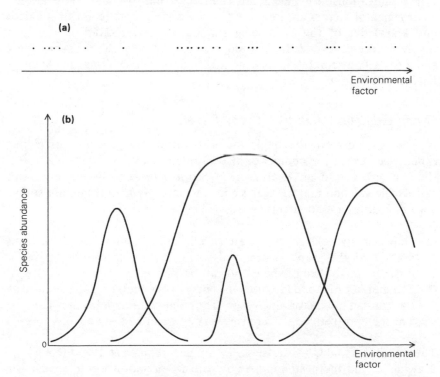

Figure 7.1 Direct Gradient Analysis of hypothetical data: (a) stand ordination; (b) a method of presenting the species ordination.

As already remarked, the idea of species occurrences and abundances along an environmental gradient is of fundamental importance in plant ecology, and leads directly to the problem of why particular species occur where they do. Direct Gradient Analysis is thus the jumping-off point for ecophysiological studies of individual plant species by both laboratory and field experimentation. Further discussion of this large and interesting subject is beyond the scope of this book, and you are referred to volumes such as Larcher (1980), Fitter & Hay (1987), Etherington (1982), which also introduce the vast literature of physiological ecology.

Since the method is so simple why bother with any other sort of ordination when Direct Gradient Analysis seems so ecologically relevant? The answer to this question falls rather naturally into two parts. First, it is not always easy or even possible to identify a single environmental gradient of overwhelming importance, and even if there is such an environmental gradient it is still only one factor out of an indefinitely large number. So to rely on only one environmental factor must inevitably give the uncomfortable feeling that important secondary factors may be overlooked. Secondly, if the indications are that more than one, but typically only a few, environmental factors are important (from a visual inspection of the data and knowledge of the area being studied), a Direct Gradient Analysis involving these several factors does not have the elegance or direct simplicity of intrepretation that a single factor Direct Gradient Analysis has, as will be seen.

Direct Gradient Analysis – many factors

As soon as we leave the single factor situation the practical difficulties rapidly increase as the number of incorporated factors is raised. Figure 7.2 shows a Direct Gradient Analysis of *Mercurialis perennis* in relation to soil extractable calcium and manganese in the Coed Nant Lolwyn case study. Two difficulties are apparent:

(a) Only one species can be shown on one graph, unless merely a small portion of the graph is occupied when another species occupying a different small area of the graph can be shown.
(b) The number of stands in different parts of the graph will be smaller as the number of factors increases. Thus the contour lines are liable to become approximate, and only a very few can be shown.

Three factors could be shown on a single graph, but the difficulty of drawing it, and the near impossibility of interpretation would render the exercise very dubious. There are, in fact, two broad categories of Direct Gradient Analysis involving many factors: direct use of the factors them-

148

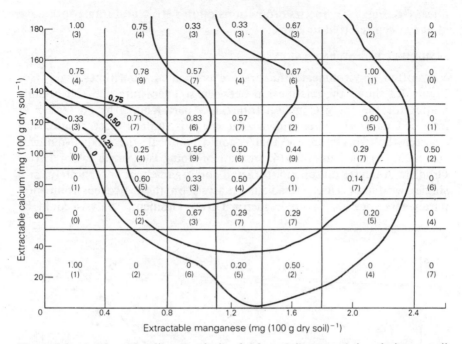

Figure 7.2 A Direct Gradient Analysis of *Mercurialis perennis* in relation to soil extractable calcium and manganese in the Coed Nant Lolwyn case study. The plane produced by the two environmental factor axes has been divided up into a grid, with each rectangular area defining a range of combined levels of calcium and manganese. Within each area, the bracketed number is the number of sample stands occurring in that range of soil calcium and manganese levels, and the other number is the proportion of those stands containing *Mercurialis perennis*. Approximate contour lines enclose areas containing proportional frequencies of this species of about 0.75, 0.5, 0.25, and enclosing an area of non-occurrence.

selves, which we may term **Direct Gradient Analysis** *sensu stricto*; and use of combinations of the environmental factors as axes, which can be referred to as **Semi-direct Gradient Analysis.**

DIRECT GRADIENT ANALYSIS *SENSU STRICTO*

The concept here is as straightforward as in the single factor case, but now the stand ordination is not completely trivial. In fact, the stand ordination is important because of the possibility of correlation between two or more environmental factors. In the case of (say) a pair of factors, a correlation coefficient can be calculated, but a graph of the two factors may be more informative in case the relationship between them is curvilinear rather than

149

linear (see Fig. 7.4). Species ordinations on this structural framework have already been referred to above.

SEMI-DIRECT GRADIENT ANALYSIS

Very often, environmental factors are correlated, sometimes highly so. Figure 7.3 shows the relationship between soil moisture and soil organic matter for the Iping Common transect. Apart from a broad scatter at intermediate levels, it is evident that there is quite a close relationship between the two factors and that it could be linear; the product-moment correlation coefficient is 0.914. Instead of doing two single factor Direct Gradient Analyses on soil moisture and soil organic matter separately, or a single two-factor Direct Gradient Analysis on these two environmental parameters together, a single factor Direct Gradient Analysis could be

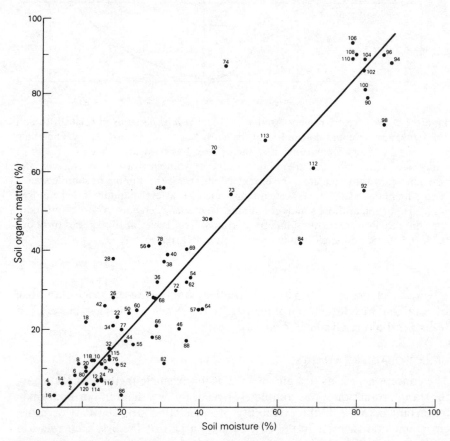

Figure 7.3 The relationship between soil moisture and soil organic matter (a stand Direct Gradient Analysis) in the Iping Common case study.

150

carried out using the diagonal straight line as the factor. This line is the best **linear combination** between these two soil quantities and stand scores are obtained along this line by constructing perpendiculars from each stand onto the line.

A similar situation exists between height of ground and soil moisture content in the same data set (Fig. 7.4), but here the relationship is obviously curvilinear, and there is vastly more scatter at the low soil moisture end. It is much more difficult to use a curve as combined factor gradient and indeed, to my knowledge, it has not been attempted. However, in this case a curvilinear relationship is clearly superior to a linear one: the product-moment correlation coefficient between soil moisture and height of ground, which assumes a linear relationship is -0.59; but the correlation coefficient

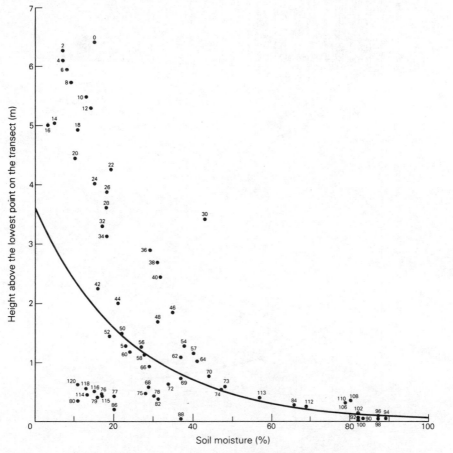

Figure 7.4 The relationship between soil moisture and topographical height (a stand Direct Gradient Analysis) in the Iping Common case study.

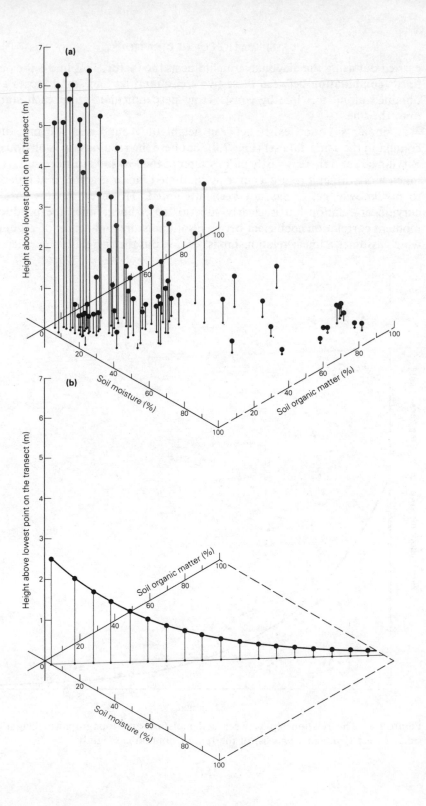

between soil moisture and the logarithm of height, which gives the exponential curve shown in Figure 7.4, is −0.74.

Consideration of the three factors of soil moisture, organic matter and ground height together gives a three-dimensional graph (Fig. 7.5a), and the relationship between this graph and the two-dimensional Figures 7.3 and 7.4 is obvious. The thick curve in Figure 7.5b over the linear relationship between soil moisture and organic matter could, in theory, be used as the basis of a single factor Direct Gradient Analysis incorporating all three environmental factors.

Using these ideas, a practical Semi-direct Gradient Analysis is effected through the method of **Principal Component Analysis** which constructs linear axes through 'clouds' of points in as many dimensions as necessary. Considering the environmental factors of soil moisture, soil organic matter, soil pH, height above ground and vegetation height of the Iping Common data, a Principal Component Analysis is shown in Figure 7.11a (stand ordination). The first axis is a linear combination of all five environmental factors, but soil moisture and organic matter have by far the greatest influence on the nature of this axis. The second axis is almost entirely a vegetation height component (Fig. 7.11b). A species ordination based on a similar analysis is shown in Fig. 9.22.

Indirect Gradient Analysis

A glance at Figures 7.3 and 7.11a enables you to compare the results of a two-factor Direct Gradient Analysis *sensu stricto* with a two-factor Semi-direct Gradient Analysis, the latter in fact incorporating five environmental factors. It is evident that the Direct Gradient Analysis *sensu stricto* is better, in so far as the environmental factors are explicitly defined.

However, the big disadvantage of Direct Gradient Analysis *sensu stricto* is the very limited number of factors that can be dealt with in one analysis. In the case of Semi-direct Gradient Analysis, while the idea of combining several environmental factors into fewer more fundamental components by Principal Component Analysis is a good one because of the existence of correlations among environmental factors, the basis of Principal Component Analysis is mathematical. *Plants, however, do not respond to environmental factors by simple mathematical rules; plants respond to combinations of environmental factors differently from the way in which*

Figure 7.5 The relationship between soil moisture, soil organic matter and topographical height (a stand Direct Gradient Analysis) in the Iping Common case study: (a) the individual stand positions; (b) a possible curvilinear theoretical relationship between the three factors (thick curve) with a projected linear relationship between soil moisture and soil organic matter (thin line).

Principal Component Analysis combines the factors. This sentence provides the rationale for indirect gradient analyses, or ordinations in the strict sense, namely that we require meaningful *biological* combinations of environmental gradients obtained *indirectly* from vegetational gradients. Vegetational gradients are obtained by ordinating the vegetation data rather than the environmental data, but after having done this we then have the problem of interpreting the vegetation gradients in environmental terms.

Indirect Gradient Analysis – one factor (Continuum Analysis)

The methods of Indirect Gradient Analysis are almost wholly applied to several vegetational gradients: these are ordination methods in the usual sense of the word. Only one method of a single factor Indirect Gradient Analysis exists, which is known as **Continuum Analysis**. The method was developed by Cottam (1949) and Curtis & McIntosh (1950, 1951), and applied to the vegetation of the upland forests of northern Wisconsin by Brown & Curtis (1952). A part of their results is shown in Figure 7.13a. Here the gradient is called a **continuum index,** and is obtained from the vegetation data themselves. Figures 7.13b,c & d show that the continuum index does indeed reflect environmental gradients.

Indirect Gradient Analysis – many factors

Since these ordination methods form the main subject matter of this

Table 7.1 A classification of ordination methods in terms of direct and indirect gradient analyses.

	Gradient Analysis	
	Direct	Indirect
One factor	essentially there is only one method, but there are many ways of depicting the results	'Continuum Analysis'

Many factors	(a) using the individual environmental factors – Direct Gradient Analysis *sensu stricto*
	 (b) Using combinations of the environmental factors – Semi-direct Gradient Analysis
		Ordination *sensu stricto*
		 (a) polar
		  Bray & Curtis (1957)
		  Orloci (1966)
		 (b) non-polar
		  Principal Component
		  Reciprocal Averaging
		  Detrended
		  Correspondence

Plate 1 A grid of strings, suitable for finding randomly chosen positions in a study area.

Plate 2 A frame of 'pins' (point quadrats).

Plate 3 Using a frame of 'pins' (point quadrats).

Plate 4 Using the point-centred quarter method in chalk grassland: measuring the distance from a randomly chosen point, marked by the centre of the cross, to a plant of *Blackstonia perfoliata*.

chapter, nothing further need be said at this point. A summary of this classification of ordinations is given in Table 7.1.

Direct Gradient Analysis – one factor

To obtain smoothed species abundance curves in relation to an environmental gradient, such as appear in Figure 7.1b, requires two steps: first, the specification of abundances of each species at particular levels of the environmental factor, or over particular ranges of levels of the environmental factor, to give a graph like Figure 9.19; and secondly to smooth the data, either by freehand curve drawing or by fitting a mathematical relationship to the points. If the latter is done it enormously complicates an otherwise elementary process; and if the former smoothing method is applied, it is subjective, with different people drawing different curves for the same data. Very often, smoothing is unnecessary if reasonably regular data are obtained as in Figure 9.19.

The kind of abundance value that is suitable depends partly on the scale of the survey. Some Direct Gradient Analyses have been applied on a very large scale, such as altitude. In one particular study, frequency estimates of each species were made at each height above sea level sampled, and the resulting graph was quite regular. For small-scale surveys the problem is more difficult; the common abundance values do not usually give a very clear picture. For example, in the Iping Common data, a plot of percentage cover estimates of *Calluna vulgaris* in each stand, defined by a square quadrat of area 0.25 m^2, against soil moisture gives the rather unhelpful graph of Figure 7.6a. About the only things one can glean from this graph is that *Calluna vulgaris* does not occur at soil moisture levels above 50%, and that in the driest soils ($<10\%$ moisture) the species always has cover values in excess of 40%. On the other hand, Figure 7.6b shows a much clearer picture from the same data set, in which the percentage frequency of occurrence of *Calluna vulgaris* among the stands in given ranges of soil moistures is used as the abundance value.

We can summarize by pointing out that in a large-scale survey the stands can be very large (several square metres), especially if the environmental factor of interest is coarse-grained, i.e. does not vary greatly over short distances. Hence, randomly sub-sampling within each stand to obtain a frequency value is appropriate (local frequency, see p.20). For small-scale surveys, such as form the case studies of this book, not only is each stand small but we are also concerned with small-scale environmental heterogeneity. Hence a large stand would be both vegetationally and environmentally heterogeneous. To overcome this difficulty, the environmental factor concerned is divided into ranges of values.

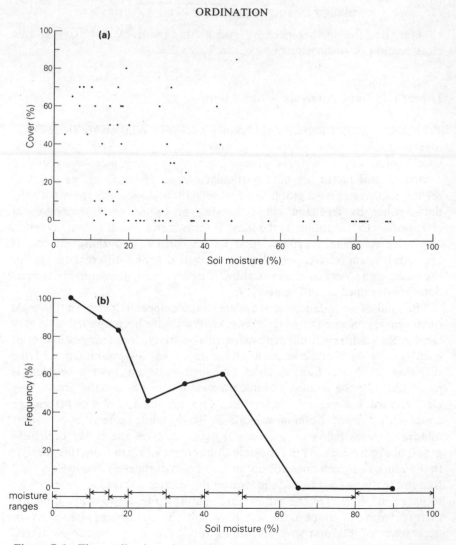

Figure 7.6 The application of two kinds of abundance value in a Direct Gradient Analysis of *Calluna vulgaris* in relation to soil moisture of the Iping Common case study: (a) using estimated percentage cover values in each stand; (b) using percentage frequency of occurrence in stands over the indicated ranges of soil moisture.

Division of the environmental gradient into ranges is rather subjective, but two guidelines may be given. First, the number of ranges must not be so large that too few stands occur within any one range. Secondly, it is expedient to aim for equality of stand number within each range, as the precision of the frequency values is thereby equalized. This is the reason for the disparity of the ranges in Figure 7.6b, from 5% to 30%. The total number of stands within each range can be found by inspection of Figure 7.6a for this data set.

Despite the apparent chaos of Figure 7.6a, Figure 7.6b shows a clear pattern of descreasing *Calluna vulgaris* frequency with increasing soil moisture. Once again, the superiority of essentially presence/absence data is demonstrated, because this is what frequency data really are.

Direct Gradient Analysis – many factors

As already discussed, there are two methods of Direct Gradient Analysis involving several factors – fully direct and semi-direct.

Direct Gradient Analysis sensu stricto

With several factors, relevant pairs can be graphed (Fig. 7.2), but it is often difficult to decide from inspection of the data which are the important pairs of factors. Moreover, with small-scale data, where frequency in ranges is the abundance value used, the number of combinations of ranges is greatly increased on a two factor graph. This reduces the number of stands in each range combination, which tends to make the frequency values irregular.

Semi-direct Gradient Analysis – Principal Component Analysis

Principal Component Analysis (PCA) is the most usual way of combining a number of environmental factors into fewer uncorrelated components. Mathematically, it is quite an involved method, using several concepts in matrix algebra. We shall not describe the mathematics of the method in any detail here since this can be found in other books, e.g. Causton (1987), but we must make some attempt to understand the salient features in relation to just two factors in geometric terms.

Figure 7.7a shows a hypothetical situation of two environmental factors, x_1 and x_2, and they are evidently quite highly correlated. Figure 7.7b shows the same data now centred on the origin; that is, the centroid of the cloud of points is at the origin. This is achieved by subtracting the means of the two variables from their respective values for each point. Note that the *structure* of the data is preserved: the points still have the same distances and orientations in relation to one another, and the correlation between the two variates is unchanged.

Having centred the data (but this is not an absolute necessity to the method of PCA), the axes are now rotated about the origin to new positions such that the first axis, y_1, lies along the most elongated direction of the data or, to put it another way, so that the first axis lies in the direction of

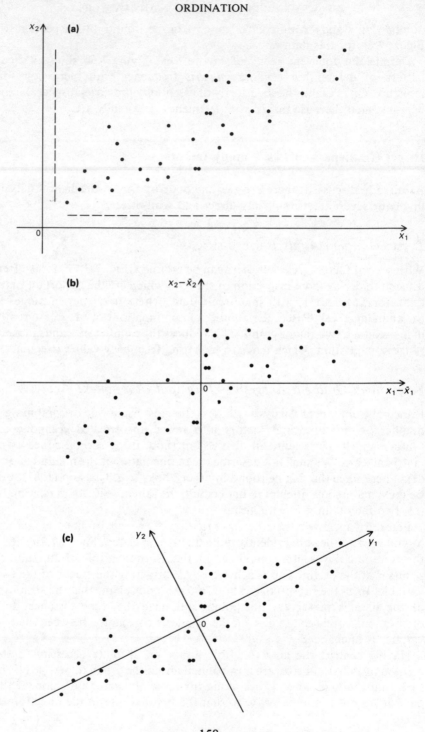

maximum variation. Whereas for the unrotated axes, the variances of the two environmental factors are of a similar order (crudely, the magnitude of a variance is reflected by the length of axis occupied by projections onto it from the extrema of a swarm of points – the dashed lines in Fig. 7.7a), for the rotated axes the variance of y_1 is large and that of y_2 is small. Rotation of the axes also preserves the structure of the data and, most importantly, rotation also preserves the *total variability*. This means that the sum of the variances of the y_1 and y_2 is equal to the sum of the variances of x_1 and x_2. However, the variability has been *repartitioned*; and further, while the factors x_1 and x_2 are highly correlated, y_1 and y_2 are uncorrelated. The rotated axes, y_1 and y_2, are known as **components,** and there are as many components as there are original factors in the data set. The question naturally arises as to the nature of the components – they are, in fact, linear combinations of the original factors, i.e.

$$y_1 = x_1 \cdot \cos \alpha + x_2 \cdot \sin \alpha$$

$$y_2 = -x_1 \cdot \sin \alpha + x_2 \cdot \cos \alpha$$

where α is the angle of rotation. This pair of linear equations is usefully put into matrix form:

$$\mathbf{y} = \mathbf{Ax} \tag{7.1}$$

where

$$\mathbf{y} = \begin{bmatrix} y_1 \\ y_2 \end{bmatrix}, \ \mathbf{x} = \begin{bmatrix} x_1 \\ x_2 \end{bmatrix}, \ \mathbf{A} = \begin{bmatrix} \cos \alpha & \sin \alpha \\ -\sin \alpha & \cos \alpha \end{bmatrix}$$

The matrix $\mathbf{A}$ is known as the **transformation matrix,** its rows are called **eigenvectors** or **characteristic vectors,** and its elements contain vital information about the nature of the components. Figure 7.8 shows two contrasting situations. In Figure 7.8a the angle of rotation between x_1 and y_1 is small; evidently there is not much difference in the nature of x_1 and y_1, and we can say that y_1 is very similar to x_1. In Figure 7.8b the angle of rotation between x_1 and y_1 is large; now there is obviously little correspondence between this factor and this component, and so x_1 contributes only a small amount of information towards y_1.

Figure 7.7 The basis of Principal Component Analysis for two environmental factors: (a) hypothetical data plotted against environmental factor axes; (b) the same data 'centred' on the origin; (c) the same data with rotated axes. The hatched lines in (a), parallel to the co-ordinate axes, show the ranges of levels of the environmental factors in the data.

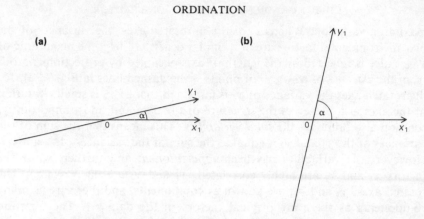

Figure 7.8 Illustrating loadings of original factors on new components: (a) factor x_1 with high loading on component y_1 – small angle of rotation; (b) factor x_1 with low loading on component y_1 – large angle of rotation.

The elements of the transformation matrix are cosines of the angles of rotation. That $\sin \alpha$ elements can also be regarded as cosines is demonstrated by Figure 7.9 and the basic trigonometrical relationships

$$\beta = 90° - \alpha$$

$$\cos(90° - \alpha) = \sin \alpha$$

hence $\cos \beta = \sin \alpha$. Therefore the transformation matrix can be written in the form

$$\mathbf{A} = \begin{bmatrix} \cos \alpha & \cos \beta \\ -\cos \beta & \cos \alpha \end{bmatrix}$$

Now the cosine of a small angle is large, and vice versa; also a small angle between a factor and a component implies a high loading between the two, so the element in the matrix designating this loading is large. The relationship between factor and component in a transformation matrix is shown below:

	Factor 1	Factor 2
Component y_1	$\cos \alpha$	$\cos \beta$
Component y_2	$-\cos \beta$	$\cos \alpha$

so the rows of the transformation matrix – the eigenvectors – correspond to

160

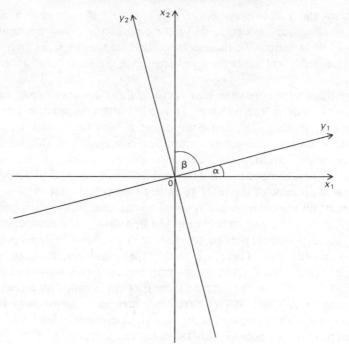

Figure 7.9 Principal Component Analysis of two factors showing the angles of rotation. For explanation, see text.

the components, and the columns of the matrix correspond to the original environmental factors.

One final feature which must be discussed concerns positive and negative elements in the transformation matrix, that is, positive and negative loadings. Figure 7.10 shows two kinds for an equal amount of loading in the absolute sense. Figure 7.10a shows a positive loading where the factor and the component both point in the same direction; in Figure 7.10b the

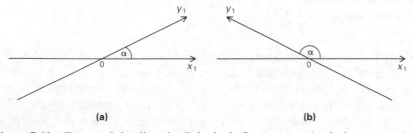

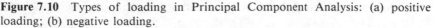

Figure 7.10 Types of loading in Principal Component Analysis: (a) positive loading; (b) negative loading.

161

component and factor are pointing in opposite directions – a negative loading – although the degree of loading of factor on component is the same in both instances. In the interpretation of loadings the sign may or may not be important depending on the precise question being asked at the time.

Where there are more than two factors, there are more than two new components; indeed, the numbers of factors and components are always equal to one another. The first component always lies in the direction of maximum variation, the second component always lies in the direction of the next highest variation but subject to the constraint that it is at right angles to the first component, and so on. Each component thus accounts for a decreasing amount of the total variation (remember that the sum of the variances of all the components is equal to the sum of the variances of all the original factors), and so it is only the first few of the components that are of great importance, because together they contain a high proportion of the total information. These are termed **principal components**, but the number of these is a subjective decision to be taken each time an analysis is performed. The variance of each new component is called an **eigenvalue** or **characteristic root**, and the set of eigenvalues and eigenvectors together constitute the main results of a Principal Component Analysis. All the foregoing theory can now be illustrated by an example.

Example 7.1

We shall use a subset of the Iping Common environmental data – those factors other than soil extractable ions – namely:

(a) soil pH,
(b) soil moisture (%),
(c) height of ground above the lowest point on the transect (m),
(d) height of vegetation (cm), and
(e) soil organic matter (%).

Principal Component Analysis works not directly on the data themselves – the level of each environmental factor in each stand – but on a summarization of the data known as the variance–covariance matrix. This is a symmetric matrix having the variances of the environmental factors in the leading diagonal, and the covariances elsewhere. Since the matrix is symmetric we need only quote the leading diagonal and the elements on one side of it. For our present set of data, the variance–covariance matrix is

$$
\begin{bmatrix}
0.0724 & & & & \\
-4.0566 & 627.6490 & & & \\
0.3233 & -28.8227 & 3.7687 & & \\
-1.8944 & 182.2018 & -20.4075 & 751.6151 & \\
-3.7510 & 631.7733 & -26.5275 & 165.3882 & 761.2185
\end{bmatrix}
$$

The positions of the elements in this matrix reflect the order of the environmental factors in the list above. Thus, 627.6490 is the variance of soil moisture, and -26.5275 is the covariance between soil organic matter and height of ground. The total of the variances is 2144.3237 which is a measure of the total variation of the data.

The results of the Principal Component Analysis are summarized by the eigenvalues (the variances of the components) and the eigenvectors (the loadings of the environmental factors on the new components), and are shown in Table 7.2. Since the first two components between them account for over 97% of the total variation, we regard these as the principal

Table 7.2 Results of the Principal Component Analysis applied to the variance–covariance matrix of the five environmental factors – soil pH, soil moisture, height of ground, vegetation height, soil organic matter – of the Iping Common data.

Eigenvalues:					Total
1420.8930	662.8228	58.3727	2.1971	0.0381	2144.3237

Percentage of total:					
66.2630	30.9106	2.7222	0.1025	0.0018	

Cumulative percentage:					
66.2630	97.1736	99.8958	99.9982	100.0000	

Transformation matrix (rows are eigenvectors):

-0.0041	0.6324	-0.0308	0.3444	0.6932
0.0001	-0.2019	-0.0089	0.9378	-0.2821
-0.0082	0.7454	-0.0567	-0.0394	-0.6629
0.0546	0.0605	0.9964	0.0167	-0.0189
0.9985	0.0055	-0.0550	0.0000	-0.0016

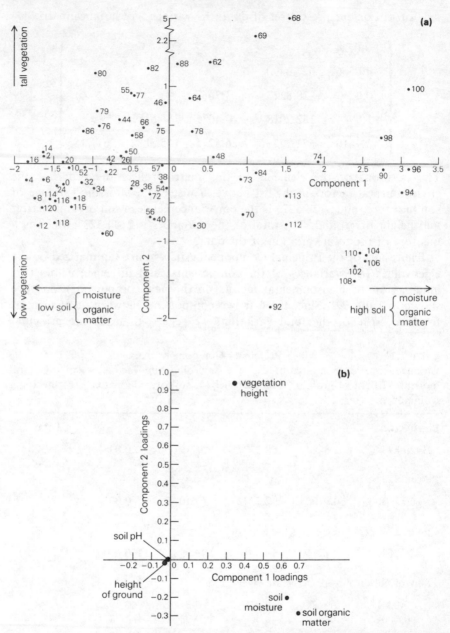

Figure 7.11 Principal Component Analysis (a semi-direct Gradient Analysis), based on the variance–covariance matrix of five environmental factors of the Iping Common case study: (a) stand ordination; (b) factor loadings. See text, Example 7.1, for further details.

components. This is convenient since it means we can summarize 97% of the total information furnished by these data on two-dimensional graphs.

To interpret the principal components, we turn to the first two rows of the transformation matrix – the first two eigenvectors. We find that two environmental factors have moderately high loadings on the first component, namely, soil organic matter and soil moisture. That both these factors 'go together' in this way should cause us no surprise as we have already seen that they are very highly correlated. Since both loadings are positive, it implies that a stand having a high score on component 1 tends to have high soil organic matter and high moisture levels, and vice versa. The second component has a different interpretation because here we have one outstandingly high loading – vegetation height; thus component 2 can be regarded simply as nearly synonymous with vegetation height. Again the loading is positive, which implies tall vegetation in stands at the high positive end of the axis, and short vegetation at the high negative end of the second axis. The results, in the form of a graph of the stands in relation to the first two components, are shown in Figure 7.11a. This is, in effect, a stand ordination based on the environmental data. Graphing the elements of the first two eigenvectors gives an environmental factor ordination (Fig. 7.11b). For example, the position of soil moisture is given by plotting the point whose co ordinates are (0.6324, − 0.2019), the second element in each of the first two eigenvectors.

To obtain a species ordination, one may proceed in the same way as for a two factor Direct Gradient Analysis *sensu stricto*, and results are shown in Figure 9.20 for *Calluna vulgaris,* and Figure 9.21 for both the lichen species and *Sphagnum recurvum.* Alternatively, one can use the centroid method of obtaining a species ordination based on a stand ordination (see p. 239) and achieve a result like that shown in Figure 9.22.

STANDARDIZED DATA

One of the disdvantages of Principal Component Analysis is that factors of high variability will tend to dominate the analysis. This means that environmental factors having high variances will have high loadings on the first component. The chief aim of the Principal Component Analysis is to reduce the 'dimensionality' of the data by extracting a few important components which together contain most of the total variation or information in the data set. The loadings of the original factors on the new components then give an idea of the importance of each factor, in that those factors highly loaded on the first, most important, component must then be regarded as leading factors in the system under study. Now one way in which a factor can have a high loading on component 1 is by having an outstandingly high variance. But from a data structure point of view, a factor is not necessarily important simply because it has a high variance; it is the nature and sizes of the correlations between the factors which are important.

In Example 7.1, we found that soil organic matter, soil moisture, and vegetation height are the highly loaded factors on the first two components. However, we know that the first two of these are also quite highly correlated with soil pH and height of ground; yet the loadings of these latter two factors on the first component are very small: why? The answer is given by examining the variances, and it is evident that the ranking of environmental factor variance size is exactly duplicated in the ranking of absolute (sign ignored) loading size on the first component. So even though a factor may be important, it will fail to be adequately expressed if its variability is low.

In view of this, it is natural to seek a way of equalizing the variances while keeping the correlation structure of the data intact. This is known as data **standardization**, and can be achieved by dividing each factor value in a stand by the standard deviation of that factor. The resulting variance–covariance matrix is now actually a matrix of correlation coefficients with the elements of the leading diagonal, the variances, all unity, thus demonstrating the standardizing effect. Rather than going through the standardizing process for each number in the data matrix, we can achieve the same effect by carrying through a Principal Component Analysis on the correlation matrix. The next example does this for the same data that were considered in Example 7.1.

Example 7.2

The correlation matrix for the five environmental factors considered in Example 7.1 is

$$
\begin{bmatrix}
1 & & & & \\
-0.6018 & 1 & & & \\
0.6190 & -0.5926 & 1 & & \\
-0.2568 & 0.2653 & -0.3834 & 1 & \\
-0.5053 & 0.9140 & -0.4953 & 0.2187 & 1
\end{bmatrix}
$$

and the eigenvalues and eigenvectors of this matrix, the results of the Principal Component Analysis, are given in Table 7.3. The first component accounts for over 60% of the total variation, not all that much less than for component 1 in the analysis of the variance–covariance matrix; but the pattern of environmental factor loadings is quite different. Here, four out of the five factors have roughly equal loadings on the first component – soil moisture, soil organic matter, height of ground and soil pH. Soil pH and

Table 7.3 Results of the Principal Component Analysis applied to the correlation matrix of the five environmental factors – soil pH, soil moisture, height of ground, vegetation height, soil organic matter – of the Iping Common data.

Eigenvalues:					Total
3.0285	0.9143	0.6177	0.3641	0.0754	5.000

Percentage of total:

60.5698	18.2870	12.3543	7.2814	1.5075

Cumulative percentage:

60.5698	78.8568	91.2111	98.4925	100.0000

Transformation matrix (rows are eigenvectors):

0.4514	− 0.5237	0.4604	− 0.2654	− 0.4894
0.0050	− 0.2850	− 0.2117	0.8636	− 0.3579
0.6111	0.2931	0.4350	0.3863	0.4497
− 0.6456	− 0.0017	0.7407	0.1857	0.0024
− 0.0771	− 0.7474	− 0.0730	0.0077	0.6580

height of ground are positively loaded, and soil moisture and organic matter are negatively loaded; this accords with the types of correlation (see correlation matrix above).

The second component accounts for a substantially lower percentage of the remaining variation (some 18%) than in the case of the variance–covariance analysis (nearly 31%), but the loading pattern is similar in that vegetation height has a very high loading on component 2 compared with the other factors. The sum of squares of the loadings on one component, i.e. the elements in an eigenvector, is always unity. Hence, if there is one outstandingly high loading on a component, the remainder must be small; conversely, if there are a number of factors of roughly equal importance, as in the first component of this example, they must all have intermediate loading values.

The stand ordination of the present analysis is shown in Figure 7.12a. It will be seen as very similar to the graph derived from the analysis based on the variance–covariance matrix. The environmental factor ordination, however, (Fig. 7.12b) will be seen as rather different from that produced by the previous analysis.

The two analyses presented in Examples 7.1 and 7.2 of the same data sets give similar results. This is by no means always the case, and we need to enquire why these data give similar results when analyzed both ways. The

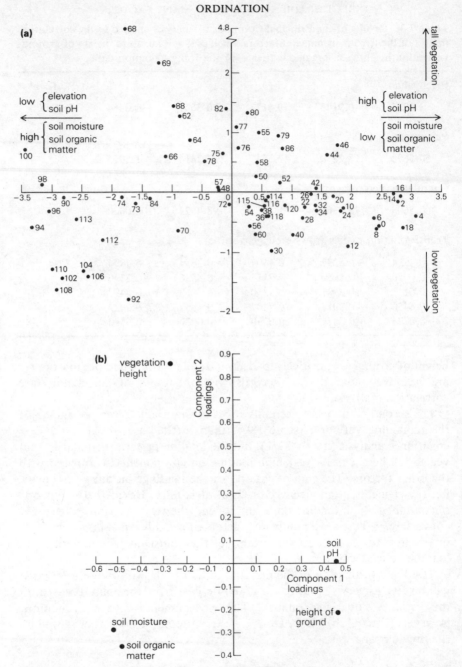

Figure 7.12 Principal Component Analysis based on the correlation matrix of five environmental factors of the Iping Common case study: (a) stand ordination; (b) factor loadings. See text, Example 7.2, for further details.

reason is that the environmental factor having the highest variance, soil organic matter, also happens to be an important one from the point of view of being highly correlated with soil moisture, which in turn is moderately correlated with soil pH and height above ground. Thus four out of the five environmental factors form a fundamental component – the first – of the data, which accounts for a large proportion of the total variation. From a data structure viewpoint, both soil pH and ground height should have a high loading on component 1 of the variance–covariance PCA, but this cannot be expressed because of their small variances. When all the factors are measured on an equal basis, i.e. when they are standardized (PCA carried out on the correlation matrix), then their true loadings are manifest.

Indirect Gradient Analysis – one factor

In their large-scale study of the upland forests of northern Wisconsin, Brown & Curtis (1952) were faced with the problem of how to arrange their sample stands into some kind of meaningful order. The basic data consisted of importance values for the tree species calculated by the formulae on pages 57–8, and for each stand the leading dominant was defined as that species with the highest importance value. The stands were then re-arranged into groups where all the stands in a group had the same leading dominant.

Stands dominated by *Acer saccharum* (sugar maple) are considered to be climax woodland in Wisconsin; and *Pinus banksiana* (jack pine), which has an importance value of zero in *Acer saccharum* woodland, is a pioneer species. By inspection of the summarized data, and from ecological knowledge, each tree species was then assigned a *climax adaptation number* on an arbitrary scale of 1 (*Pinus banksiana*) to 10 (*Acer saccharum*). Finally, each stand is assigned a continuum index value, given by

$$c_i = \sum_{j-1}^{s} v_{ij} a_j$$

where c_i is the continuum index of the ith stand, s is the total number of species, v_{ij} is the importance value of species j in the ith stand, and a_j is the climax adaptation number of the jth species. A fuller description of the method is given in Kershaw & Looney (1985).

The continuum index is regarded as a measure of the total environment of a stand expressed in terms of species composition and relative species abundance. This is why the method is one of Indirect Gradient Analysis. But having 'integrated the environment' into one factor – the continuum index – we may now plot the results, in the form of a species ordination, as in the one factor Direct Gradient Analysis; and the results for some species,

169

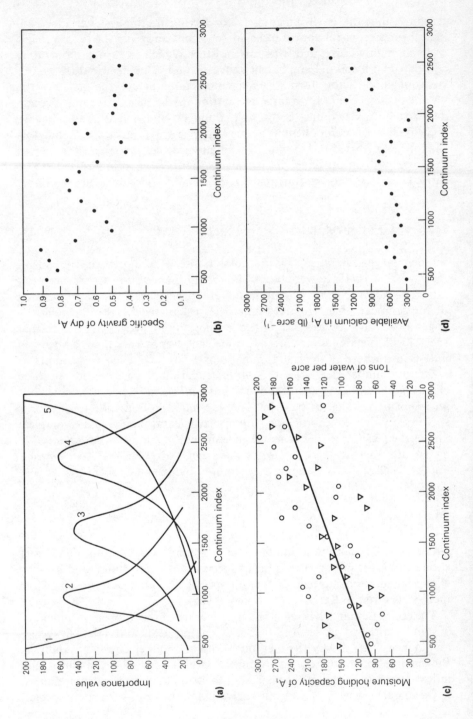

170

in the form of smoothed curves, is shown in Figure 7.13a. That the continuum index is indeed a reflection of environmental changes is shown by plots of three soil factors in the A₁ horizon against stand continuum index in Figure 7.13b,c,d: definite trends are discernible.

Although this method, known as **Continuum Analysis**, has its points of interest, it has been little used subsequently. This is probably due to the fact that a subjective judgement is required at one stage of the process, and that with the development of automatic high speed electronic computing, several efficient and entirely objective methods of ordination (Indirect Gradient Analysis involving many axes, not just one) have been developed. This is why Continuum Analysis has not been considered in detail here.

Indirect Gradient Analysis – many factors (ordination *sensu stricto*)

The rest of this chapter is concerned with the ordination methods which have been generally applied to vegetation data. In terms of our present framework, ordination methods can all be viewed as indirect gradient analyses involving more than one vegetational gradient. Before embarking on detailed descriptions and discussions of the methods, we must first be aware that they are divisible into two broad categories: **polar** and **non-polar** methods.

The distinction between polar and non-polar ordination methods is shown in Figure 7.14. Both graphs show the same data – a number of stands in a two-species space – but the method of establishing an ordination axis differs. In any ordination method, the aim is to establish axes in directions of maximum variation (see discussion on Principal Component Analysis as a semi-direct gradient analysis on p. 157), but there is no unique way of doing this. In Figure 7.14a an axis is placed along the direction of maximum variation with regard to the cluster of data points as a whole – non-polar ordination; whereas in Figure 7.14b, the axis is placed as a line joining the two stands which are furthest apart – polar ordination.

In the situation shown in Figure 7.14, which could be typical of some practical data sets, it would seem that non-polar ordination would be the

---

Figure 7.13 Some results of a Continuum Analysis of stands from upland forests of northern Wisconsin: (a) species importance values plotted against continuum index for (1) *Pinus banksiana* (Jack pine), (2) *Pinus resinosa* (red pine), (3) *Pinus strobus* (white pine), (4) *Tsuga canadensis* (eastern hemlock), (5) *Acer saccharum* (sugar maple); (b) specific gravity of soil samples from the A₁ horizon of stands grouped into classes of 100-unit widths of continuum index; (c) moisture holding capacity of the A₁ horizon, grouped as in (b); (d) available calcium in the A₁ horizon, grouped as in (b). (After Brown & Curtis 1952, by permission of *Ecological monographs*.)

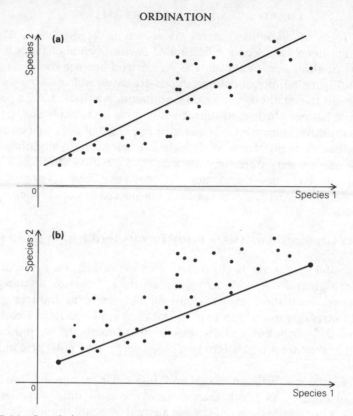

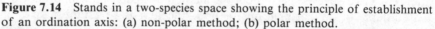

Figure 7.14 Stands in a two-species space showing the principle of establishment of an ordination axis: (a) non-polar method; (b) polar method.

more efficient since it is assessing the data as a whole rather than two individual stands; clearly, the axis in Figure 7.14b is not the best one in relation to the whole data set.

Polar methods of ordination were developed before non-polar ones, partly at least because the former are computationally simpler than the latter. Furthermore, polar methods have been developed specifically for vegetation ordinations, whereas non-polar methods are mostly statistical methods of quite general utility which, as one of their many uses, have been employed as ordination procedures. This is certainly true for Principal Component Analysis and, although apparently developed specifically for vegetation ordination purposes, Reciprocal Averaging and its derivative – Detrended Correspondence Analysis – are essentially more general techniques, but so far have been little used elsewhere. Further discussion of the merits and disadvantages of the two ordination systems is best deferred until their methodology and application to real data sets have been considered.

172

Polar ordination

Polar ordination methods establish their axes by joining the most dissimilar stands, or species, at each stage. We shall examine two methods: Bray & Curtis' original procedure, and Orloci's later modification of it.

The ordination of Bray & Curtis (1957)

In the recent past, the Bray & Curtis method of ordination has been regarded as of historical interest only. However, because the method performed quite well in a study using artificial data whose underlying structure was known (see p. 214); and also because the method is quite well suited to the modern interactive computer scenario, especially to microcomputers where one may not be able to mount a large program required by non-polar ordination methods, interest in the method is reviving. Bray & Curtis' ordination has been successfully used in the past and, being simple, it highlights the problems of polar ordination quite well. Moreover, once the initial calculations of dissimilarities between each pair of stands has been done, say on a microcomputer, the ordination could be constructed using nothing more than ruler and compasses.

The first task is to calculate a dissimilarity index between each pair of stands. In their prototype, Bray & Curtis used the Czekanowski coefficient on quantitative data as the similarity index, and then calculated dissimilarities as

$$D_C = 1 - S_C$$

For qualitative data, either the Jaccard or the Sorensen coefficients are equivalent.

The two most dissimilar stands are selected, and a straight line joining these two stands forms the first axis; the score of one of these stands on the first axis is aribitrarily defined as zero, and the other stand then has the score D_{max} on the first axis (Fig. 7.15). The scores of all the other stands on Axis 1 are obtained as shown in Figure 7.15, where the d-values are the appropriate dissimilarity coefficients. For a direct geometrical construction, the positions of x_i can be obtained by compass constructions of the S_i and then dropping a perpendicular from S_i onto the first axis (to T_i); this is shown in Figure 7.16a. Alternatively, for computation we have the following derivation.

In triangle $S_1 S_i T_i$, Pythagoras' theorem gives

$$h_i^2 = d_{1i}^2 - x_{1i}^2 \tag{7.2}$$

173

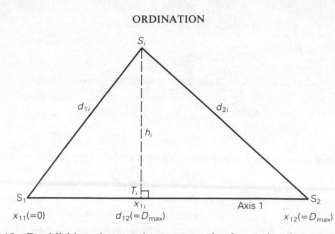

Figure 7.15 Establishing the stand scores on the first axis of a Bray & Curtis Ordination. One of the two reference stands (S_1, say) is given the score of zero, and the other reference stand (S_2, say) then has score D_{max} on the first axis. The remaining stands ($S_i, i = 3, ..., n$; n = total number of stands) have scores on the final axis of x_{1i}, calculated as shown in the text.

For triangle $S_2S_iT_i$, the same theorem gives

$$h_i^2 = d_{2i}^2 - (d_{12} - x_{1i})^2 \qquad (7.3)$$

Multiplying out the bracket in (7.3), and eliminating h_i^2 between the two equations gives

$$d_{1i}^2 - x_{1i}^2 = d_{2i}^2 - d_{12}^2 + 2d_{12}x_{1i} - x_{1i}^2$$

The x_{1i}^2 terms cancel, and re-arrangement finally gives

$$x_{1i} = \frac{d_{1i}^2 - d_{2i}^2 + d_{12}^2}{2d_{12}} \qquad (7.4)$$

so the score of the ith stand on Axis 1, x_{1i}, is given wholly in terms of the original dissimilarities.

A second axis is formed by joining a pair of stands which have similar scores on Axis 1, but also have a large dissimilarity between themselves. Almost always the second axis will be oblique to the first, so if there is any

Figure 7.16 Stages in the construction of the Bray & Curtis Ordination of stands of the quantitative Artificial Data: (a) showing the compass method of positioning the points representing the stands, and their projection onto the first axis (cf. Fig. 7.15); (b) the stand ordination on the first axis; (c) the two-dimensional ordination with oblique axes.

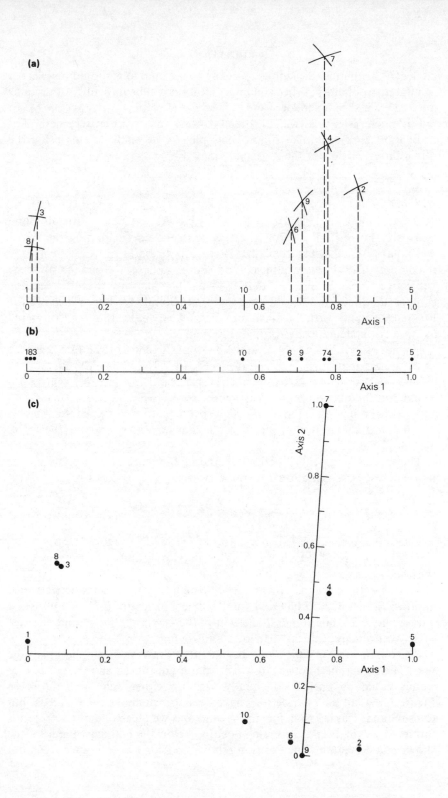

choice in the matter, the stands to define the second axis should be selected to minimize obliquity; otherwise interpretation will be difficult. A third axis may be erected in a similar way.

It is much easier to discuss the details of this, rather subjective, method of ordination by means of a simple example; so we shall do this using the quantitative version of the Artificial Data.

Example 7.3

The calculations and results of the Czekanowski coefficient for all stand pairs are shown in Table 7.4, together with the corresponding matrix of dissimilarity coefficients beneath. There are 16 pairs of stands with the maximum dissimilarity coefficient of unity, and so the choice must be subjective. If these were real data it would be valid and useful to employ ecological criteria; but for now we shall take Stands 1 and 5, which are the first pair of stands with the maximum dissimilarity index that is encountered as we work systematically through the matrix.

Having established Stands 1 and 5 at opposite ends of the first axis, the other stands may now be positioned. Figure 7.16a shows the method of geometrical construction. Arcs of circles of radii d_{1i} and d_{5i}, centred on Stand 1 at 0 units and Stand 5 at 1 unit, respectively, are constructed and from their point of intersection a perpendicular is dropped to Axis 1. The point of intersection of the perpendicular with Axis 1 defines the score, x_{1i}.

The second, and better, method of finding the score of each stand on the first axis is to use Equation 7.4 which, with $d_{12} = 1$, reduces to

$$x_{1i} = (d_{1i}^2 - d_{2i}^2 + 1)/2$$

The results are given in Table 7.5, and are plotted in Figure 7.16b.

To obtain a second axis, we examine pairs of stands which occur close together on Axis 1 to see which of these have large dissimilarity values. None of the three stands 1, 3 and 8 have high distance values between one another; Stands 4 and 7 have a dissimilarity coefficient of 0.71, Stands 6 and 9 have only 0.27, but Stands 7 and 9 have the maximum interstand distance of 1. Accordingly, this last pair of stands define Axis 2.

There are two ways which have been used to draw Axis 2 in relation to the first. The earlier method was to work from Figure 7.16a and recognize that stands could be positioned equally well on either side of Axis 1. In Figure 7.16a, all the intersections have been drawn above the first axis, but to use Stands 7 and 9 to define the second axis we place one of these stands (arbitrarily 9) in its corresponding position below the first axis. Figure 7.16c shows Stand 7 in the same position relative to Axis 1 as in Figure 7.16a but

Table 7.4 Czekanowski similarity coefficients, S_C, for all pairs of stands of the Artificial Data (quantitative); and the matrix of dissimilarity coefficients $D_C(=1-S_C)$. Notation as Equation 5.17.

Stand pair	$\Sigma x_j + \Sigma y_j$	$2\Sigma \min(x_j, y_j)$	S_C	Stand pair	$\Sigma x_j + \Sigma y_j$	$2\Sigma \min(x_j, y_j)$	S_C
1,2	23	2	0.09	3,10	26	12	0.46
1,3	23	18	0.78	4,5	15	8	0.53
1,4	17	2	0.12	4,6	14	6	0.43
1,5	22	0	0	4,7	14	4	0.29
1,6	21	6	0.29	4,8	14	0	0
1,7	21	0	0	4,9	18	6	0.33
1,8	21	18	0.86	4,10	20	4	0.20
1,9	25	6	0.24	5,6	19	12	0.63
1,10	27	12	0.44	5,7	19	6	0.32
2,3	22	0	0	5,8	19	0	0
2,4	16	6	0.38	5,9	23	14	0.61
2,5	21	14	0.67	5,10	25	14	0.56
2,6	20	14	0.70	6.7	18	0	0
2,7	20	0	0	6,8	18	2	0.11
2,8	20	0	0	6,9	22	16	0.73
2,9	24	18	0.75	6,10	24	14	0.58
2,10	26	18	0.69	7,8	18	0	0
3,4	16	0	0	7,9	22	0	0
3,5	21	0	0	7,10	24	0	0
3,6	20	2	0.10	8,9	22	0	0
3,7	20	0	0	8,10	24	12	0.50
3,8	20	18	0.90	9,10	28	16	0.57
3,9	24	0	0				

	1	2	3	4	5	6	7	8	9	10
1	—									
2	0.91	—								
3	0.22	1.00	—							
4	0.88	0.64	1.00	—						
5	1.00	0.33	1.00	0.47	—					
6	0.71	0.30	0.90	0.57	0.37	—				
7	1.00	1.00	1.00	0.71	0.68	1.00	—			
8	0.14	1.00	0.10	1.00	1.00	0.89	1.00	—		
9	0.76	0.25	1.00	0.67	0.39	0.27	1.00	1.00	—	
10	0.56	0.31	0.54	0.80	0.44	0.42	1.00	0.50	0.43	—

with Stand 9 in its proper location below Axis 1. The main disadvantage of this procedure is that the second axis is oblique to the first, and so the plotting of the stands and the interpretation of their positions is not easy. In fact, the two axes are not **orthogonal,** which means that any gradients represented by each of the two axes are not independent. In constructing

Table 7.5 Calculations of the Bray & Curtis stand ordination. Axis 1 scores of the stands of the quantitative Artificial Data, assuming $d_{15} = 1$, $x_{11} = 0$, $x_{15} = 1$.

Stand (i)	d_{1i}	d_{5i}	x_{1i}
1	0	1.00	0
2	0.91	0.33	0.86
3	0.22	1.00	0.02
4	0.88	0.47	0.78
5	1.00	0	1.00
6	0.71	0.37	0.68
7	1.00	0.68	0.77
8	0.14	1.00	0.01
9	0.76	0.39	0.71
10	0.56	0.44	0.56

Figure 7.16c, each point has co-ordinates given by the final columns in Tables 7.5 and 7.6, and the points are plotted at the positions of intersection of perpendiculars to each axis at the appropriate scores.

Figure 7.17 shows the same ordination with non-oblique, or orthogonal, axes. This graph has been constructed simply by drawing the two axes at right angles and plotting the co-ordinate values of each stand in the usual way.

It would be very difficult using the first method to extend the ordination

Table 7.6 Calculations of the Bray & Curtis stand ordination. Axis 2 scores of the stands of the quantitative Artificial Data, assuming $d_{79} = 1$, $x_{29} = 0$, $x_{27} = 1$.

Stand (i)	d_{9i}	d_{7i}	x_{2i}
1	0.76	1.00	0.29
2	0.25	1.00	0.03
3	1.00	1.00	0.50
4	0.67	0.71	0.47
5	0.39	0.68	0.34
6	0.27	1.00	0.04
7	1.00	0	1.00
8	1.00	1.00	0.50
9	0	1.00	0
10	0.43	1.00	0.09

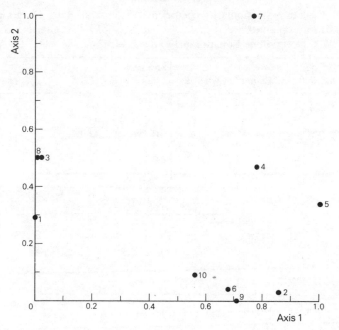

Figure 7.17 The Bray & Curtis stand ordination of the quantitative Artificial Data, with orthogonal axes.

to a third axis, but quite straightforward via the second approach. All we need to do is to find two stands in Figure 7.17 which are close together, yet have a high dissimilarity coefficient. Then the scores in the third axis are obtained in the same way as scores on the second axis where Stands 7 and 9 are at its extremes. However, for the Artificial Data no third axis is necessary since all the stands occurring close together in Figure 7.17 have small interstand distance coefficients.

Example 7.4

The Bray & Curtis species ordination of the quantitative Artificial Data can be dealt with much more briefly. The details are shown in Tables 7.7 and 7.8, and the result is shown in Figure 7.18 with orthogonal axes.

Examples 7.3 and 7.4 demonstrate two features, the first of which is common to all ordinations and the second to polar methods in particular. First, there is the problem of distortion of interstand distances. Figure 7.16a demonstrates the distortion of distances when they are projected into a space of fewer dimensions. To position three points accurately with respect to one another requires two dimensions of space, but when these distances

179

Table 7.7 Czekanowski similarity coefficients, S_C, for all pairs of species of the Artificial Data (quantitative); and the matrix of dissimilarity coefficients $D_C(= 1 - S_C)$. Notation as Equation 5.17 and Table 7.4.

Species pair	$\Sigma x_i + \Sigma y_i$	$2\Sigma \min(x_i, y_i)$	S_C	Species pair	$\Sigma x_i + \Sigma y_i$	$2\Sigma \min(x_i, y_i)$	S_C
I,II	54	18	0.33	II,IV	56	10	0.18
I,III	47	8	0.17	II,V	43	0	0
I,IV	26	2	0.08	III,IV	49	0	0
I,V	13	0	0	III,V	36	2	0.06
II,III	77	14	0.18	IV,V	15	0	0

	I	II	III	IV	V
I	—				
II	0.67	—			
III	0.83	0.82	—		
IV	0.92	0.82	1.00	—	
V	1.00	1.00	0.94	1.00	—

Table 7.8 Calculations of the Axis 1 and Axis 2 scores of the Bray & Curtis species ordination of the Artificial Data (quantitative), assuming $d_{I,IV} = 1$, $x_{1II} = 0$, $x_{1V} = 1$ for the first axis; $d_{III,IV} = 1$, $x_{2III} = 0$, $x_{2IV} = 1$ for the second axis.

First axis

Species	d_{Ij}	d_{Vj}	x_{1j}
I	0	1.00	0
II	0.67	1.00	0.23
III	0.83	0.94	0.40
IV	0.92	1.00	0.42
V	1.00	0	1.00

Second axis

Species	d_{IIIj}	d_{IVj}	x_{2j}
I	0.83	0.92	0.42
II	0.82	0.82	0.50
III	0	1.00	0
IV	1.00	0	1.00
V	0.94	1.00	0.44

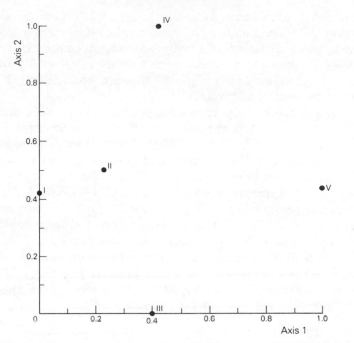

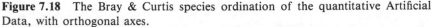

Figure 7.18 The Bray & Curtis species ordination of the quantitative Artificial Data, with orthogonal axes.

are projected into one dimension – onto Axis 1 – the distances are foreshortened. However, there is some measure of proportionality between the real and shortened distances, and this is the basis of the efficacy of ordination.

Secondly, there is no correspondence between an axis of the stand ordination (Fig. 7.17) and the corresponding axis of the species ordination (Fig. 7.18). If there was, Species V should occur near to Stand 3, since this species occurs only in this stand. On the other hand, there is nothing absolute in the direction of an axis; reversal of Axis 1 in (say) the species ordination would indeed bring Species V and Stand 3 near to one another. Since, however, the bulk of the occurrence and abundance of Species III is in Stands 1, 3, 8, one would expect this species to occur in a similar position to the stands if there was correspondence of the axes; even axis reversals would not help here.

The method of Orloci (1966)

Apart from the minor difficulty of non-orthogonal axes, which can be overcome, the Bray & Curtis' ordination has the other disadvantage of a rather imprecise definition of axes after the first. While these difficulties can be surmounted when computing by hand or when computing interactively, they can create difficulties for fully automatic computation. Orloci's method of ordination is based on similar principles to Bray & Curtis', but uses a more mathematically rigorous distance coefficient and a fully objective definition of higher axes.

Euclidean distance is the coefficient employed in Orloci's ordination; using this, the definition of the first axis and of the stand scores thereon is the same as for Bray & Curtis' method (Fig. 7.19a), so the score of the ith stand on Axis 1 is given by x_{1i}, calculated using Equation 7.4. The second axis is established by dropping a perpendicular from the stand giving the largest h-value (assumed to be Stand 3 in Figure 7.19b) to the first axis. The second axis is therefore at right angles to the first, but it is normally drawn parallel to the h_{max} line such that it intersects axis 1 at zero. Point S_1 thus becomes the origin of the co-ordinate system (see Fig. 7.21).

For point S_i to be in its correct position with respect to S_1, S_2 and S_3, it cannot in general lie in the (S_1, S_2, S_3) plane (Fig. 7.19c); but we are particularly interested in distance AB $(= x_{2i})$, the score of the ith stand on Axis 2. This is derived as follows. In triangle ADS_i

$$u_i^2 = h_i^2 + (x_{13} - x_{1i})^2 \tag{7.5}$$

All the quantities on the right-hand side are already known, so u_i can be calculated. Then, in triangle ABS_i

$$(BS_i)^2 = u_i^2 - x_{2i}^2 \tag{7.6}$$

and in triangle BS_3S_i

$$(BS_i)^2 = d_{3i}^2 - (h_{3(max)} - x_{2i})^2 \tag{7.7}$$

By eliminating $(BS_i)^2$ between Equations 7.6 and 7.7, we obtain

$$x_{2i} = \frac{u_i^2 - d_{3i}^2 + h_{3(max)}^2}{2h_{3(max)}} \tag{7.8}$$

in an analogous fashion to the derivation of Equation 7.4. Equations 7.4 and 7.8 give the scores of the ith stand on the first two axes.

The distance of point S_i from the (S_1, S_2, S_3) plane becomes important in relation to the third axis. This distance, CS_i $(= q_i)$, in relation to Axis 3 is analogous to distance DS_i $(= h_i)$ with respect to Axis 2. In the triangle CDS_1 (formed by joining C with S_1)

$$CS_1^2 = x_{1i}^2 + x_{2i}^2 \tag{7.9}$$

182

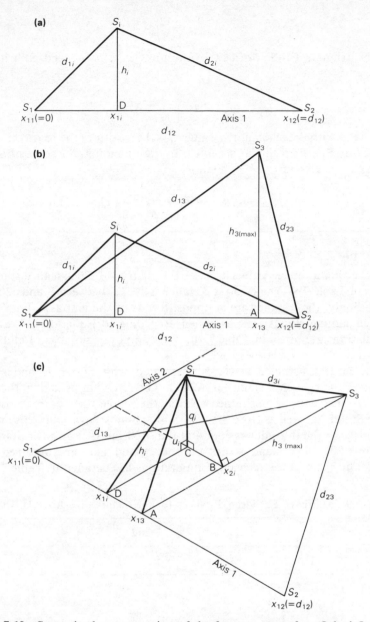

Figure 7.19 Stages in the construction of the first two axes of an Orloci Ordination: (a) the positioning of the ith stand in relation to the first axis, assuming Stands 1 and 2 are the most dissimilar and therefore define Axis 1; (b) the definition of the second axis, assuming Stand 3 has h_{max}; (c) as (b), but Stand i is raised above the (S_1, S_2, S_3) plane so that d_{3i} (also d_{1i} and d_{2i}) assume their correct lengths. The symbols d_{ij} denote distances between Stands i and j, while x_{k1} denote the score of Stand 1 on axis k. Bold lines denote those which do not lie in the horizontal plane formed by the first two axes.

183

In the triangle CS_1S_i (formed by joining C with S_1, and S_1 with S_i), $CS_i^2 = S_1S_i^2 - CS_1^2$, i.e.

$$q_i^2 = d_{1i}^2 - (x_{1i}^2 + x_{2i}^2) \qquad (7.10)$$

The stand with the maximum q-value (q_{max}) constitutes the fourth reference stand (say S_4). By a similar argument to the foregoing, the score of Stand i on Axis 3 is given by

$$x_{3i} = q_{max} - \sqrt{\{d_{4i} - (x_{1i} - x_{14})^2 - (x_{2i} - x_{24})^2\}} \qquad (7.11)$$

Example 7.5

For the quantitative Artificial Data, the matrix of Euclidean distances is given in Table 7.9. The greatest distance is 14.21, between Stands 2 and 3; accordingly, these stands are at opposite ends of the first axis (Fig. 7.20a). The remaining stands are positioned according to Equation 7.4, and the calculations are given in Table 7.10. Two stand positionings, 4 and 7, are shown in Figure 7.20a.

The largest h-value, 11.46, is associated with Stand 7, and so this becomes the reference stand for Axis 2. The remaining stands (other than 2, 3, 7) must now be positioned so that they have their correct distances from Stand 7 as well as from Stands 2 and 3; almost invariably this involves moving into the third dimension, and the situation is shown for Stand 4 in Figure 7.20b. The stand scores on the second axis are obtained from Equation 7.8, and the relevant computations are detailed in Table 7.10.

Table 7.9 Matrix of Euclidean distances for the quantitative Artificial Data.

						Stand					
		1	2	3	4	5	6	7	8	9	10
	1	—									
	2	13.60	—								
	3	3.32	14.21	—							
	4	9.64	8.25	10.49	—						
Stand	5	12.17	4.36	12.61	5.20	—					
	6	10.05	4.24	11.05	4.69	3.87	—				
	7	13.08	13.49	13.49	7.35	9.22	11.05	—			
	8	3.00	13.49	1.41	9.49	11.79	10.20	12.73	—		
	9	12.21	4.47	13.78	7.48	5.92	3.74	13.04	13.04	—	
	10	9.95	6.16	9.90	9.49	7.00	6.16	14.07	9.49	7.87	—

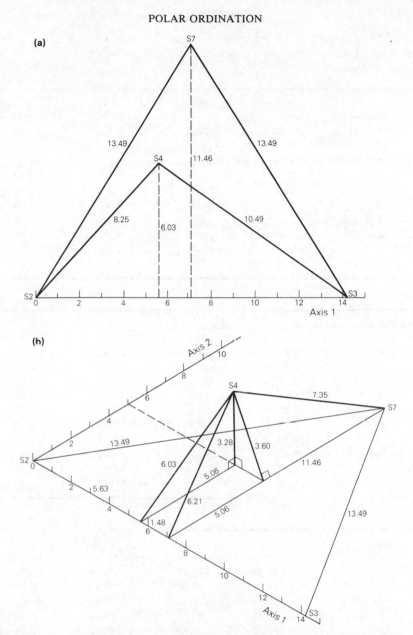

Figure 7.20 Stages in the construction of the first two axes of the Orloci Ordination of the stand quantitative Artificial Data: (a) establishment of the first axis with Stands 2 and 3, and showing the placing of Stands 7 (h_{max}) and 4 thereon; (b) showing the three-dimensional construction necessary when placing stands correctly with respect to Stand 7, which defines the end-point of Axis 2, as well as with respect to Stands 2 and 3 which define the ends of Axis 1.

Table 7.10 Calculations of the Axis 1, 2 and 3 scores for the Orloci stand ordination of the Artificial Data (quantitative).

Axis 1

Stand (i)	d_{2i}	d_{3i}	h_i	x_{1i}
1	13.60	3.32	0.61	13.23
2	0	14.21	0	0
3	14.21	0	0	14.21
4	8.25	10.49	6.03	5.63
5	4.36	12.61	3.78	2.18
6	4.24	11.05	2.48	3.44
7	13.49	13.49	11.46	7.11
8	13.49	1.41	1.16	13.44
9	4.47	13.78	4.33	1.13
10	6.16	9.90	3.61	4.99

Axis 2

Stand (i)	h_i	$x_{1i} - x_{17}$	u_i	x_{2i}
1	0.37	37.45	37.82	−0.08
2	0	50.55	50.55	0
3	0	50.41	50.41	−0.01
4	36.36	2.19	38.55	5.06
5	14.29	24.31	38.60	3.71
6	6.15	13.47	19.62	1.26
7	131.33	0	131.33	11.46
8	1.35	40.07	41.42	0.47
9	18.75	35.76	54.51	0.69
10	13.03	4.49	17.52	−2.14

Axis 3

Stand (i)	q_i	d_{9i}	$(x_{1i} - x_{19})^2$	$(x_{2i} - x_{29})^2$	x_{3i}
1	3.15	149.08	146.41	0.59	2.83
2	0	19.98	1.28	0.48	0
3	0	189.89	171.09	0.49	−0.01
4	3.28	55.95	20.25	19.10	0.20
5	0.70	35.05	1.10	9.12	−0.71
6	2.14	13.99	5.34	0.33	1.39
7	0.31	170.04	35.76	115.99	−0.01
8	1.06	170.04	151.54	0.05	−0.03
9	4.27	0	0	0	4.27
10	2.91	61.94	14.90	8.01	−3.34

The height of each stand position above the Axis 1–2 plane, q_i, is given in the first column of the Axis 3 score calculations in Table 7.10. Stand 9 has the maximum q-value of 4.27, and so forms the reference stand for the third axis. The scores of the other stands are computed from Equation 7.11, and the details are also given in Table 7.10.

Finally, it will be observed that a number of the scores on Axes 2 and 3 are negative. Such scores are perfectly valid, and if one is plotting a two-dimensional graph there is no disadvantage in having the origin elsewhere than in the bottom left-hand corner. However for a three-dimensional graph, where some of the points will lie beneath the horizontal plane formed by the first two axes, the display is much improved by arranging for all scores to be positive. This is easily achieved by adding to each stand score on a particular axis the absolute value of the highest negative score. In our present example these are -2.14 for Axis 2 and -3.34 for Axis 3, and these adjusted values have been used in the construction of the final ordination diagram shown in Figure 7.21.

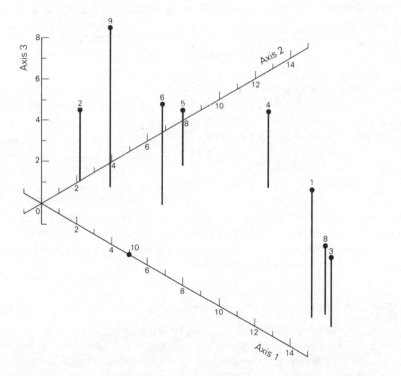

Figure 7.21 The Orloci stand ordination of the quantitative artificial data, with three axes.

187

The corresponding species ordination, and the stand and species ordinations for the qualitative Artificial Data, are left as exercises for you.

When Orloci's method of ordination was introduced, 9 years after Bray & Curtis' method, it seemed to be a real advancement in that here was a polar ordination method which was completely automatic from a computation point of view. Since most computer systems at that time did not allow user interaction while the program was running, the Orloci method was ideal for electronic computation whereas the Bray & Curtis method was not.

Non-polar ordination

Instead of establishing axes based on individual pairs of very dissimilar stands, non-polar ordination methods erect axes by consideration of directions of maximum variation of the stand set as a whole. The classical way of doing this is by PCA. First used in this context by Goodall (1954), but under the erroneous name of Factor Analysis which is related in some ways to PCA, the method was undoubtedly popularized much later in vegetation analysis by Orloci's (1966) paper; and, coupled with the increased availability of both digital computers and package programs, PCA dominated ordination methods for over a decade. However, theoretical studies (e.g. Swan 1970) highlighted the shortcomings of PCA, and towards the end of the 1970s PCA was rapidly eclipsed by the related, but somewhat more efficient, method of **Reciprocal Averaging** (RA), first described by Hill (1973). An alternative name for RA is **Correspondence Analysis,** and this contributes to the name of the newest technique of the set – **Detrended Correspondence Analysis** (DCA) (Hill 1979a, Hill & Gauch 1980) – which overcomes the major disadvantage of RA, and indeed of most other ordinations too.

The three techniques, PCA, RA and DCA, thus form a developmental sequence in methodology; they are also the predominant ordination methods employed (for almost 20 years in the case of PCA). Other non-polar ordinations have been developed and tried (see Gauch 1982) but they offer no advantages over the above trio; consequently, we shall confine attention to this triumvirate of ordination methods.

Principal Component Analysis

Principal Component Analysis is applied to vegetation data as an ordination method in precisely the same way as described already, on pages 157–62, when employed to ordinate stands on the basis of their environmental attributes. In the present case, plant species replace environmental factors, and can be represented in the data matrix either qualitatively or quantitatively. The underlying geometrical model is stands in species space; so the scores of each stand on each component, or axis, give a stand ordination, and the species loadings on each component can be graphed to give a species ordination. This is known as an R-technique. Hence, although the geometric model would seem to be only capable of yielding a stand ordination, the elements of the relevant eigenvectors give a perfectly valid species ordination simultaneously.

Example 7.6

The results of Principal Component Ordinations of the Artifical Data, starting from: (a) the variance–covariance matrix of species occurrences,

Table 7.11 Results of the Principal Component Ordination applied to the species variance–covariance matrix of the quantitative Artificial Data.

Eigenvalues:

257.3000	135.6000	31.3900	10.0900	0.6211

Percentage of total:

59.1493	31.1723	7.2161	2.3195	0.1428

Cumulative percentage:

59.1493	90.3216	97.5377	99.8572	100.0000

Transformation matrix (rows are eigenvectors: species loadings):

−0.0755	−0.6449	0.7559	−0.0775	0.0305
0.1126	0.5735	0.4296	−0.6883	0.0090
0.7669	−0.4000	−0.3046	−0.3983	−0.0217
0.6259	0.3079	0.3884	0.6009	−0.0393
0.0426	0.0180	−0.0183	0.0236	0.9985

Stand scores on first two components			Stand loadings on first two components		
1	6.84	1.12	1	0.4261	0.0961
2	−6.27	2.76	2	−0.3906	0.2373
3	7.85	1.22	3	0.4893	0.1048
4	−1.26	−3.20	4	−0.0786	−0.2749
5	−4.49	−1.14	5	−0.2798	−0.0975
6	−3.01	1.01	6	−0.1874	0.0868
7	−0.94	−9.28	7	−0.0273	−0.7969
8	7.06	0.78	8	0.4403	0.0671
9	−5.28	2.07	9	−0.3290	0.1774
10	−1.01	4.65	10	−0.0630	0.3997

Table 7.12 Results of the Principal Component Ordination applied to the species correlation matrix of the quantitative Artificial Data.

Eigenvalues:

| 2.0690 | 1.5100 | 0.7839 | 0.5243 | 0.1120 |

Percentage of total:

| 41.3866 | 30.2048 | 15.6805 | 10.4877 | 2.2404 |

Cumulative percentage:

| 41.3866 | 71.5915 | 87.2720 | 97.7596 | 100.0000 |

Transformation matrix (rows are eigenvectors: species loadings):

-0.3100	-0.5064	0.5934	-0.0772	0.5380
0.4694	0.2933	0.2883	-0.7724	0.1177
0.7133	-0.6245	0.0829	0.1906	-0.2409
0.3480	0.0948	-0.4346	0.2070	0.7989
0.2316	0.5085	0.6074	0.5642	0.0231

Stand scores on first two components			Stand loadings on first two components		
1	0.2606	0.3011	1	0.1812	0.2450
2	-0.4222	0.1540	2	-0.2935	0.1253
3	1.0561	0.1685	3	0.7342	0.1372
4	-0.1173	-0.2120	4	-0.0816	-0.1725
5	-0.2668	-0.2775	5	-0.1855	-0.2258
6	-0.2774	0.1750	6	-0.1929	0.1424
7	-0.0380	-0.9783	7	-0.0265	-0.7962
8	0.4444	0.0228	8	0.3089	0.0185
9	-0.5866	0.4784	9	-0.4078	0.3893
10	-0.0527	0.1680	10	-0.0366	0.1368

and (b) the correlation matrix of species occurrences, are given in Tables 7.11 and 7.12, respectively. In both of these analyses the eigenvectors are the species loadings and, by using Equation 7.1, the stand scores are computed and given in the bottom left-hand corner of each table. Ignore the figures in the bottom right-hand corner of the tables for the moment.

Examining the first two eigenvectors in Table 7.11, we see that Species III has a high positive loading on the first component, and Species II has a high negative loading on this component; the remaining species have negligible loadings. On Axis 2, Species II and III have fairly high positive loadings, and Species IV has a high negative loading. These features are shown in

190

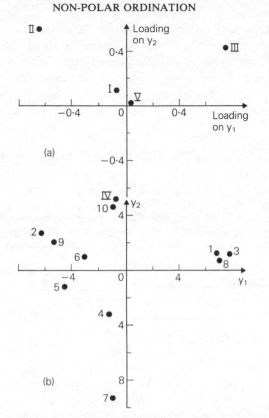

Figure 7.22 Principal Component Ordination (R-type) of the variance–covariance matrix of the Artificial Data: (a) species loadings; (b) stand scores.

Figure 7.22a. The stand scores on the first two axes are plotted in Figure 7.22b. Compare these ordinations with the results for polar ordinations earlier in this chapter.

The diagrams are not given for the correlation matrix ordination, but it is instructive to compare the species loadings (Table 7.12) with those of the variance–covariance matrix ordination (Table 7.11). While Species II and III still have high loadings at opposite ends of Axis 1 in the correlation matrix ordination, Species V also has a high loading. A glance at the data matrix shows Species V to occur in one stand only with a very low abundance value. This precludes it from showing any indication of its possible true importance in the variance–covariance matrix ordination but, in the correlation matrix ordination where each species is given equal weighting, its importance is able to be expressed (see discussions on pp. 165–6 & 201–2).

191

One could also carry out a PCA on the data matrix inversely. Now the geometric model is species in stands space, the scores of each species on each axis give a species ordination, and the elements of the relevant eigenvectors give a simultaneous stand ordination. This is known as the Q-technique.

Example 7.7

In Table 7.13 are shown the results of the Q-type Principal Component Ordination (PCO) on the variance–covariance matrix of the Artificial Data. Now the eigenvectors are the stand loadings, giving a stand ordination, and

Table 7.13 Results of the Principal Component Ordination applied to the stand variance–covariance matrix of the quantitative Artificial Data.

Eigenvalues:

258.9000	204.8000	59.8400	14.8600	0	0	0	0	0	0

Percentage of total:

48.0869	38.0386	11.1144	2.7600	0	0	0	0	0	0

Cumulative percentage:

48.0869	86.1256	97.2400	100	100	100	100	100	100	100

Transformation matrix (rows are eigenvectors: stand loadings):

0.38	−0.45	0.43	−0.09	−0.32	−0.23	−0.01	0.39	−0.37	−0.15
0.31	0.34	0.36	−0.02	0.13	0.23	−0.35	0.34	0.23	0.55
−0.03	0.09	0.16	0.14	0.38	−0.03	0.80	0.19	−0.24	0.27
0.56	−0.13	−0.14	0.19	−0.10	0.17	0.35	0.04	0.63	−0.25
−0.16	−0.15	−0.40	−0.52	0.29	0.04	−0.03	0.64	0.14	−0.13
0	0	0	0	0	0	0	0	0	0
0	0	0	0	0	0	0	0	0	0
0	0	0	0	0	0	0	0	0	0
0	0	0	0	0	0	0	0	0	0
0	0	0	0	0	0	0	0	0	0

Species scores on first two components

I	−0.74	−2.32
II	−11.55	7.22
III	11.08	7.83
IV	−0.21	−7.94
V	1.42	−4.79

Species loadings on first two components

I	−0.0462	−0.1621
II	−0.7179	0.5044
III	0.6888	0.5474
IV	−0.0130	−0.5547
V	0.0883	−0.3350

scores on the components are those of species, which give the species ordination. Again, for the moment, ignore the figures in the bottom right-hand corner of the table.

Note that the numbers of non-zero eigenvalues and eigenvectors are not greater than the lesser of the numbers of species or stands. In general, the number of non-zero eigenvalues and eigenvectors is equal to the lesser of the numbers of species or stands, but in this example it happens that the fifth eigenvalue is zero.

So it would seem that both species and stand ordinations could be carried out in two ways, *but the results are not the same.* One reason why the results differ between the R- and Q-techniques is readily apparent by inspection of the stand scores from an R-analysis, and of the species scores from a Q-analysis. Values outside the range -1 to $+1$ are commonly seen, but the species loadings from an R-analysis and the stand loadings from a Q-analysis are constrained to lie in the range -1 to $+1$ because the sum of the squares of the loadings on any one component is unity. Hence, the species and stand ordinations from any *one* analysis are obtained from different quantities, and the situation is reversed between R-type and Q-type analyses.

Now, Equation 7.1 in relation to an R-analysis shows a transformation matrix, $\mathbf{A}$, of species loadings, and n column vectors, $\mathbf{y}$, of stand scores (there is one $\mathbf{y}$ for each stand, showing the scores on all m components). If a column vector is pre-multiplied by the diagonal matrix

$$\mathbf{\Lambda}^{-1/2} = \begin{bmatrix} 1/\sqrt{\lambda_1} & & & \\ 0 & 1/\sqrt{\lambda_2} & & \\ . & . & & \\ . & . & & \\ 0 & 0 & \dots\dots\dots & 1/\sqrt{\lambda_m} \end{bmatrix}$$

where λ_i is the ith eigenvalue, we have

$$\mathbf{z} = \mathbf{\Lambda}^{-1/2}\mathbf{y} \tag{7.12}$$

The resulting column vector $\mathbf{z}$ now contains the *stand loadings* on each component, and may be compared with the appropriate elements in the transformation matrix of the Q-analysis. For instance, in Table 7.11 the results of the transformation (Equn 7.12) is given in the bottom right-hand corner for the first two components. These values may be compared directly with the first two rows of the transformation matrix in Table 7.13.

193

Evidently, there is broad similarity between the two sets of results, particularly on the first component, but they are not identical. A similar comparison may be made between the species loadings in the bottom right-hand corner of Table 7.13 and the first two rows of the transformation matrix in Table 7.11, and similar conclusions may be drawn.

The important summarization of the above paragraph is that for both R- and Q-techniques, both species and stand ordinations may be obtained as eigenvectors, and so are amenable to direct comparison. The eigenvectors of the R- and Q-analyses *should* match (species to species and stands to stands) because *the two geometric models – stands in species space, and species in stands space – are identical; they are merely different ways of depicting what is actually a single underlying mathematical model.*

DATA CENTRING

The reason for the actual mismatch in the examples of the discussion above is that although we have performed both R- and Q-analyses on the variance–covariance matrix in each case, the analyses do not treat the data in the same way. Performing a PCA on a variance–covariance matrix implies that the data are first centred on the origin, around which the axes are rotated (see p.157). In an R-type analysis the axes are the species, and so the data are said to be **centred by species**; whereas in Q-type analysis the axes are the stands, and so the data are **centred by stands.** To clarify the situation, the treatment of a very small data matrix is made in symbolic form in the following example.

Example 7.8
Consider the data matrix

$$\mathbf{X} = \begin{bmatrix} x_{11} & x_{12} \\ x_{21} & x_{22} \\ x_{31} & x_{32} \end{bmatrix}$$

where x_{ij} denotes the amount of the jth species in the ith stand; hence there are three stands and two species. Let $\bar{x}_{.j}$ denote the mean of the jth species, and let $\bar{x}_{i.}$ denote the mean of the ith stand. So, e.g. $\bar{x}_{.1} = (x_{11} + x_{21} + x_{31})/3$, and $\bar{x}_{1.} = (x_{11} + x_{12})/2$.

Centring by species To avoid a multiplicity of symbols, $\mathbf{X}$ will still be used to denote the data matrix, but now the elements are modified

$$\mathbf{X} = \begin{bmatrix} x_{11} - \bar{x}_{.1} & x_{12} - \bar{x}_{.2} \\ x_{21} - \bar{x}_{.1} & x_{22} - \bar{x}_{.2} \\ x_{31}' - \bar{x}_{.1} & x_{32} - \bar{x}_{.2} \end{bmatrix}$$

If **X** is now pre-multiplied by its transpose, we have

$$\mathbf{X}^\mathsf{T}\mathbf{X} = \begin{bmatrix} x_{11} - \bar{x}_{.1} & x_{21} - \bar{x}_{.1} & x_{31} - \bar{x}_{.1} \\ x_{12} - \bar{x}_{.2} & x_{22} - \bar{x}_{.2} & x_{32} - \bar{x}_{.2} \end{bmatrix} \begin{bmatrix} x_{11} - \bar{x}_{.1} & x_{12} - \bar{x}_{.2} \\ x_{21} - \bar{x}_{.1} & x_{22} - \bar{x}_{.2} \\ x_{31} - \bar{x}_{.1} & x_{32} - \bar{x}_{.3} \end{bmatrix}$$

on multiplying out, the product becomes

$$\mathbf{X}^\mathsf{T}\mathbf{X} = \begin{bmatrix} \sum_{i=1}^{3} (x_{i1} - \bar{x}_{.1})^2 & \sum_{i=1}^{3} (x_{i1} - \bar{x}_{.1})(x_{i2} - \bar{x}_{.2}) \\ \sum_{i=1}^{3} (x_{i2} - \bar{x}_{.2})(x_{i1} - \bar{x}_{.1}) & \sum_{i=1}^{3} (x_{i2} - \bar{x}_{.2})^2 \end{bmatrix}$$

An element in the leading diagonal is the sum of squares of the deviations of the stand scores of a species from the mean of that species: there is one element for each species. The other elements of the matrix $\mathbf{X}^\mathsf{T}\mathbf{X}$ are the sum of cross products between species 1 and 2. If each element is divided by $(n - 1)$, the result is the species variance–covariance matrix; so we can write

$$\mathbf{S_R} = \{1/(n - 1)\}\mathbf{X}^\mathsf{T}\mathbf{X} \qquad (7.13)$$

n in this case being 3. The above is a demonstration of the fact that when PCA is applied to the species variance–covariance matrix, the data are actually centred by species in the process.

Centring by stands We now modify the data matrix as

$$\mathbf{X} = \begin{bmatrix} x_{11} - \bar{x}_{1.} & x_{12} - \bar{x}_{1.} \\ x_{21} - \bar{x}_{2.} & x_{22} - \bar{x}_{2.} \\ x_{31} - \bar{x}_{3.} & x_{32} - \bar{x}_{3.} \end{bmatrix}$$

that is, the data are centred by stands. Now we post-multiply **X** by its transpose:

$$\mathbf{XX^T} = \begin{bmatrix} x_{11} - \bar{x}_{1.} & x_{12} - \bar{x}_{1.} \\ x_{21} - \bar{x}_{2.} & x_{22} - \bar{x}_{2.} \\ x_{31} - \bar{x}_{3.} & x_{32} - \bar{x}_{3.} \end{bmatrix} \begin{bmatrix} x_{11} - \bar{x}_{1.} & x_{21} - \bar{x}_{2.} & x_{31} - \bar{x}_{3.} \\ x_{12} - \bar{x}_{1.} & x_{22} - \bar{x}_{2.} & x_{32} - \bar{x}_{3.} \end{bmatrix}$$

which yields

$$\mathbf{XX^T} = \begin{bmatrix} \sum_{j=1}^{2} (x_{1j} - \bar{x}_{1.})^2 & & \\ \sum_{j=1}^{2} (x_{2j} - \bar{x}_{2.})(x_{1j} - \bar{x}_{1.}) & \sum_{j=1}^{2} (x_{2j} - \bar{x}_{2.})^2 & \\ \sum_{j=1}^{2} (x_{3j} - \bar{x}_{3.})(x_{1j} - \bar{x}_{1.}) & \sum_{j=1}^{2} (x_{3j} - \bar{x}_{3.})(x_{2j} - \bar{x}_{2.}) & \sum_{j=1}^{2} (x_{3j} - \bar{x}_{3.})^2 \end{bmatrix}$$

where, because the matrix is symmetric, it is only partially quoted. This time, a diagonal element is the sum of squares of the deviations of the species scores in a stand from that stand's mean, and the other elements are the sums of cross products between pairs of stands. Dividing each element by $(m-1)$, we have the stand variance–covariance matrix:

$$\mathbf{S}_Q = \{1/(m-1)\}\mathbf{XX^T} \tag{7.14}$$

m in this case being 2. So if a PCA is carried out on the stands variance–covariance matrix, the data are centred by stands in the process.

Non-centred data A PCA can, in fact, be carried out on non-centred data. Using the original symbolic data matrix in this example, we now have either

$$\mathbf{X^TX} = \begin{bmatrix} x_{11} & x_{21} & x_{31} \\ x_{12} & x_{22} & x_{32} \end{bmatrix} \begin{bmatrix} x_{11} & x_{12} \\ x_{21} & x_{22} \\ x_{31} & x_{32} \end{bmatrix}$$

$$= \begin{bmatrix} x_{11}^2 + x_{21}^2 + x_{31}^2 & x_{11}x_{12} + x_{21}x_{22} + x_{31}x_{32} \\ x_{12}x_{11} + x_{22}x_{21} + x_{32}x_{31} & x_{12}^2 + x_{22}^2 + x_{32}^2 \end{bmatrix}$$

i.e.

$$\mathbf{X}^T\mathbf{X} = \begin{bmatrix} \sum\limits_{i=1}^{3} x_{i1}^2 & \sum\limits_{i=1}^{3} x_{i1}x_{i2} \\[2em] \sum\limits_{i=1}^{3} x_{i2}x_{i1} & \sum\limits_{i=1}^{3} x_{i2}^2 \end{bmatrix}$$

or

$$\mathbf{X}\mathbf{X}^T = \begin{bmatrix} x_{11} & x_{12} \\ x_{21} & x_{22} \\ x_{31} & x_{32} \end{bmatrix} \begin{bmatrix} x_{11} & x_{21} & x_{31} \\ x_{12} & x_{22} & x_{32} \end{bmatrix}$$

$$= \begin{bmatrix} x_{11}^2 + x_{12}^2 & x_{11}x_{21} + x_{12}x_{22} & x_{11}x_{31} + x_{12}x_{32} \\ x_{21}x_{11} + x_{22}x_{12} & x_{21}^2 + x_{22}^2 & x_{21}x_{31} + x_{22}x_{32} \\ x_{31}x_{11} + x_{32}x_{12} & x_{31}x_{21} + x_{32}x_{22} & x_{31}^2 + x_{32}^2 \end{bmatrix}$$

i.e.

$$\mathbf{X}\mathbf{X}^T = \begin{bmatrix} \sum\limits_{i=1}^{2} x_{1j}^2 & \sum\limits_{i=1}^{2} x_{1j}x_{2j} & \sum\limits_{i=1}^{2} x_{1j}x_{3j} \\[2em] \sum\limits_{i=1}^{2} x_{2j}x_{1j} & \sum\limits_{i=1}^{2} x_{2j}^2 & \sum\limits_{i=1}^{2} x_{2j}x_{3j} \\[2em] \sum\limits_{i=1}^{2} x_{3j}x_{1j} & \sum\limits_{i=1}^{2} x_{3j}x_{2j} & \sum\limits_{i=1}^{2} x_{3j}^2 \end{bmatrix}$$

Submitting either $\mathbf{X}^T\mathbf{X}$ or $\mathbf{X}\mathbf{X}^T$ to PCA will produce *identical* results, with just two new components (the lesser of the stands and species). The important feature here is that *for both the $\mathbf{X}^T\mathbf{X}$ and $\mathbf{X}\mathbf{X}^T$ matrices the data have been transformed in the same way*, i.e. not transformed at all! This is why submission of either matrix to PCA produces the same results.

Why carry out a non-centred PCA at all? Dagnelie (1960) seems to have been the first to try this variant of PCA in ordination, and he found that the first component comprised all positive loadings which simply reflected the general species abundances. Dagnelie therefore considered that the first axis

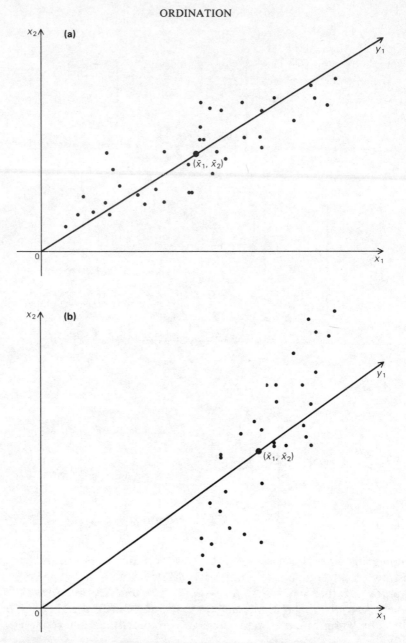

Figure 7.23 Non-centred Principal Component Ordination in two-dimensional space: (a) the hypothetical data of Figure 7.7a, in which the orientations of the first axes of centred and non-centred ordinations are similar; (b) another set of hypothetical data in which the orientations of the first axes of centred and non-centred ordinations are different.

in a non-centred PCA was not particularly informative, and so advised against this approach.

Dagnelie's findings can be appreciated by inspection of Figure 7.23, which shows the situation for just two species (stands in species space). The axes are rotated about the origin, and the first new component, y_1, is shown; the second component axis (not shown) is at right-angles to y_1, also through the origin. It is immediately evident why all loadings and scores on y_1 are positive, and that loadings on y_2 can be either positive or negative (perhaps slightly less immediately evident!). The two graphs in Figure 7.23 also show two contrasting situations. In Figure 7.23a the angle of rotation is very similar to that which would be obtained from centred data (cf. Fig. 7.7); but in Figure 7.23b the result would be very different, with a much higher proportion of the total variability accounted for by the second component in non-centred rotation. At first sight, PCA applied to the data situation of Figure 7.23b would fail to give any sort of meaningful result.

However, by using an artificial data matrix with one discontinuity in it (i.e. there were two groups of stands, with no common species between the two groups), Noy-Meir (1973) demonstrated that non-centred PCA showed this disjunction clearly, whereas any kind of centring blurred the discontinuity. In the same paper he also claimed similar trends in real field data, namely, that non-centred PCAs tended to expose clusters of distinctive stands more effectively than centred PCAs.

Although the concept is important, few people have been drawn to the method of non-centred PCA, and are unlikely to be so in the future as usage of PCA has declined in favour of Reciprocal Averaging and Detrended Correspondence ordinations. As may be seen from the results of the Coed Nant Lolwyn case study, the differences between centred and non-centred PCA are not dramatic unless there are discontinuities in the data matrix.

DATA STANDARDIZATION

Much more important than the centring/non-centring concept is that of data standardization. There are various ways of doing this, but the commonest is to divide an observation of the jth variate (species in an R-analysis, stands in a Q-analysis) by its own standard deviation. Example 7.9 shows how this is done symbolically.

Example 7.9

Centring and standardization by species   It will be more illuminating to consider both centred and standardized data. Starting with the initial data matrix of Example 7.8, centring and standardization give the following modified data matrix:

$$\mathbf{X} = \begin{bmatrix} \dfrac{x_{11} - \bar{x}_{.1}}{s_{.1}} & \dfrac{x_{12} - \bar{x}_{.2}}{s_{.2}} \\[2ex] \dfrac{x_{21} - \bar{x}_{.1}}{s_{.1}} & \dfrac{x_{22} - \bar{x}_{.2}}{s_{.2}} \\[2ex] \dfrac{x_{31} - \bar{x}_{.1}}{s_{.1}} & \dfrac{x_{32} - \bar{x}_{.2}}{s_{.2}} \end{bmatrix}$$

from which

$$\mathbf{X}^T\mathbf{X} = \begin{bmatrix} \dfrac{\sum\limits_{i=1}^{3} (x_{i1} - \bar{x}_{.1})^2}{s_{.1}^2} \\[3ex] \dfrac{\sum\limits_{i=1}^{3} (x_{i2} - \bar{x}_{.2})(x_{i1} - \bar{x}_{.1})}{s_{.2}s_{.1}} \quad \dfrac{\sum\limits_{i=1}^{3} (x_{i2} - \bar{x}_{.2})^2}{s_{.2}^2} \end{bmatrix}$$

where $s_{.j}^2$ is the variance of the jth species. Now multiply both sides by the scalar $1/(n-1)$. The numerator of (say) the first element in the leading diagonal becomes

$$\frac{\sum\limits_{i=1}^{3} (x_{i1} - \bar{x}_{.1})^2}{n-1}$$

which equals $s_{.1}^2$, the variance of species 1. Hence, the element itself is $s_{.1}^2/s_{.1}^2 = 1$; likewise, for other elements in the leading diagonal. The numerators of the off-diagonal elements become covariances on multiplying by $1/(n-1)$, i.e.

$$s_{21} = \frac{\sum\limits_{i=1}^{3} (x_{i2} - \bar{x}_{.2})(x_{i1} - \bar{x}_{.1})}{n-1}$$

and the covariance divided by the product of the standard deviations is the product-moment correlation coefficient; so

$$r_{21} = \frac{s_{21}}{s_{.2}s_{.1}}$$

Thus, we have that

$$\{1/(n-1)\}\mathbf{X}^\mathrm{T}\mathbf{X} = \begin{bmatrix} 1 & \\ r_{21} & 1 \end{bmatrix}$$

n in this case being 3. So we have the important result that the variance—covariance matrix of data standardized by standard deviations is the correlation matrix of the original (but centred) data. Hence, using the species correlation matrix in a PCA implies doing the analysis on data centred and standardized by species.

Centring and standardization by stands Here, we have

$$\{1/(m-1)\}\mathbf{X}\mathbf{X}^\mathrm{T} = \begin{bmatrix} 1 & & \\ r_{21} & 1 & \\ r_{31} & r_{32} & 1 \end{bmatrix}$$

where m in this case is 2. Hence, using the stand correlation matrix in a PCA implies doing the analysis on data centred and standardized by stands.

Standardizing the data has the same effect now that we are ordinating stands by their floristic composition as in the case of ordinating stands by their environmental make-up (p. 166). In the discussion on page 165 it was pointed out that performing a PCA on a variance—covariance matrix would cause environmental factors of high variability to be highly loaded onto the first component. The same thing happens with vegetation data and, as it is usually the case that replicate observations having a high mean usually have a high variance also and vice versa, we find that PCA carried out on the variance—covariance matrix of a vegetation data matrix shows species loadings on the first component roughly proportional to the 'amounts' of the species in the data set. Thus, with qualitative data, the species of the highest frequencies tend to have the highest loadings on the first component; for quantitative data, both a species' frequency and its abundance within the stands contribute to its loading on the first component. The effects on the first component have been stressed because it is the most important, but similar effects occur with respect to the other components also. As a summary of the content of this paragraph, we can say that using a variance—covariance matrix in a PCA weights the species according to their frequency (and abundance).

Use of a species correlation matrix (or data standardized by species) in a PCA results in each species having an equal weight. The results of a PCA now only reflect correlations of occurrence, and so tend to be more ecologically meaningful. It is for this reason that most PCA ordinations have been carried out on a standardized data matrix.

Since PCA ordinations can be done on: data centred by species, data centred by stands, or non-centred data; on data standardized by species, standardized by stands, or non-standardized data; on qualitative or quantitative data, in all combinations; one set of data can obviously generate a very large number of PCA variants. Mostly, the differences between these are quite minor, and the most usual approach is a PCA ordination using the species correlation matrix, or in other words, a PCA ordination carried out on data both centred and standardized by species. More information can be found in Noy-Meir (1973); Noy-Meir *et al.* (1975); and in an account based on these two papers by Greig-Smith (1983, pp. 247–57).

Reciprocal Averaging (Hill 1973)

The method of Reciprocal Averaging, or Correspondence Analysis to give it its alternative name, is interesting because it has affinities with both Direct Gradient Analysis and Principal Component Analysis.

THE GRADIENT ANALYSIS APPROACH

By viewing Reciprocal Averaging in relation to Gradient Analysis, the computations for the ordination can be done in an elementary way, starting by either weighting the species according to their supposed occurrence in relation to an obviously important environmental gradient, or by weighting the stands according to their level of an important environmental factor. In reality though, *any* weighting can be given at the expense of increasing the amount of computation required; but since for real data the calculations are too laborious to do by hand anyway, this feature is not crucial. Rather than try to describe the procedure in words, we shall proceed straight to an example.

Example 7.10

As before in this chapter, we shall use the quantitative Artificial Data; these are re-stated in Table 7.14. Rather than giving the species or stands arbitrary starting scores, we shall assign starting values to the stands using the Orloci Ordination stand scores on the first axis (Table 7.10). These are not copied directly, but are first converted to a 0–100 range, purely for convenience, by the formula

$$y = \frac{100(x - x_{\min})}{x_{\max} - x_{\min}} \tag{7.15}$$

Table 7.14 The quantitative Artificial Data, together with stages in the computation of the first axis of a Reciprocal Averaging Ordination.

Stand	Species					Stand totals	Column		
	I	II	III	IV	V		(1a)	(2)	(2a)
1	3	0	9	0	0	12	93	66	79
2	1	10	0	0	0	11	0	2	0
3	0	0	10	0	1	11	100	83	100
4	1	2	0	2	0	5	40	16	17
5	0	7	0	3	0	10	15	9	9
6	2	6	1	0	0	9	24	14	15
7	0	0	0	9	0	9	50	29	33
8	0	0	9	0	0	9	95	81	98
9	5	8	0	0	0	13	8	8	7
10	0	9	6	0	0	15	35	32	37
Species totals	12	42	35	14	1				

Row		I	II	III	IV	V
	(1)	34	17	84	41	100
	(1a)	21	0	81	29	100
	(2)	29	14	81	26	100
	(2a)	17	0	78	14	100

		Standardized results										Converged result
		(1a)	(2a)	(3a)	(4a)	(5a)	(6a)	(7a)	(8a)	(9a)	(10a)	
	1	93	79	78	79	81	83	86	88	89	89	91
	2	0	0	0	1	16	27	36	42	46	49	58
	3	100	100	100	100	100	100	100	100	100	100	100
	4	40	17	9	4	12	18	23	27	29	31	36
Stands	5	15	9	3	0	10	18	24	28	32	33	40
	6	24	15	14	15	28	37	44	50	54	56	64
	7	50	33	15	0	0	0	0	0	0	0	0
	8	95	98	97	98	98	98	99	99	99	99	99
	9	8	7	6	8	21	32	40	46	49	52	61
	10	35	37	37	39	48	55	61	64	66	68	74
	I	21	17	15	25	33	40	46	50	53	55	61
	II	0	0	0	12	21	30	36	40	43	44	52
Species	III	81	78	77	81	80	85	87	88	89	89	90
	IV	29	14	0	0	0	0	0	0	0	0	0
	V	100	100	100	100	100	100	100	100	100	100	100

where x is an original score, x_{min} and x_{max} are the minimum and maximum original scores, respectively, and y is the new score on the 0–100 scale. The denominator on the right-hand side of Equation 7.15 is, of course, the range of the original scores. For our present conversion, $x_{min} = 0$ and $x_{max} = 14.21$, and so $y = 100x/14.21$. Having converted the Orloci Ordination Axis 1 stand scores to the new range, they are written into Table 7.14 as Column (1a). We now calculate a first set of species scores by a weighted average technique; so for Species I, we have

$$\frac{(3)(93) + (1)(0) + (1)(40) + (2)(24) + (5)(8)}{12}$$

$$= \frac{279 + 0 + 40 + 48 + 40}{12} = \frac{407}{12} = 33.9$$

i.e. for each stand containing Species I, we multiply the abundance of the species in that stand by the stand score, and finally divide the sum of these products by the Species I total. This is done for the remaining species, and

Table 7.15 The first three axes of the Reciprocal Averaging Ordination of the quantitative Artificial Data.

	Axis		
	1	2	3
eigenvalue	0.7896	0.6119	0.1572
stand			
1	− 178	66	62
2	50	− 155	− 39
3	− 240	139	− 25
4	199	7	32
5	173	− 39	− 47
6	8	− 118	14
7	447	238	11
8	− 233	130	− 11
9	29	− 145	65
10	− 59	− 42	− 47
species			
I	− 12	− 123	286
II	56	− 158	− 71
III	− 233	130	− 11
IV	447	238	11
V	− 304	227	− 160

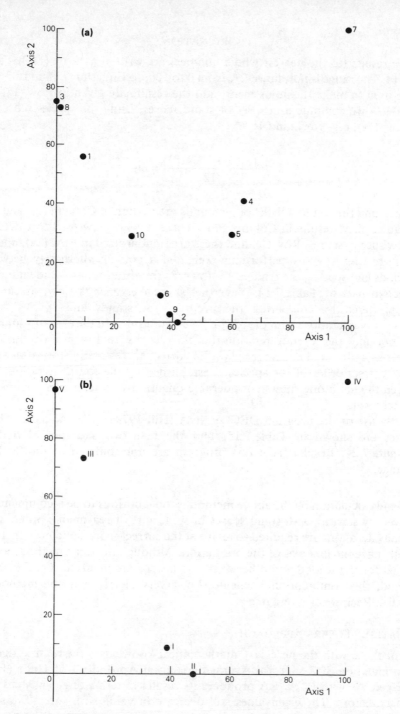

Figure 7.24 Reciprocal Averaging Ordination of the quantitative Artificial Data: (a) stands; (b) species.

the results (to the nearest whole number) are written as Row (1) in Table 7.14. The range is not, however, from 0 to 100, so Equation 7.15 must again be used to make the adjustment, and the results are given in Row (1a).

Now we compute a new set of stand scores, using the species scores of Row (1a); e.g. for Stand 1:

$$\frac{(3)(21) + (9)(81)}{12} = 66$$

This, and the results for the other stands are written in Column (2), and the standardized results in Column (2a). These last are now used to calculate new species scores, Row (2), and these are standardized in Row (2a), and so on. In this way, by performing weighted averaging alternately between stands and species, the tendency is towards stabilization, as is shown by the second part of Table 7.14. Nevertheless, convergence is slow, as can be appreciated by comparing the progress for stands and species from successive iterations with the fully converged results. In the case of stands, it is not until the seventh iteration that they have sorted themselves out into correct order, and convergence to the correct scores is still slow after that; the correct order of the species is established by the fourth iteration, but even the tenth one shows considerable quantitative discrepancies from the final result.

By use of the program DECORANA (Hill 1979a), the results for three axes are shown in Table 7.15, and the first two axes are plotted in Figure 7.24. Results from this program are not confined to the (0–100) range.

Hand calculation by the above method is too laborious to be recommended. Even by starting with stand scores in some sort of reasonable order, very many iterations are required to arrive at the correct scores; and even then we only have the first axis of the ordination. Although the same method can be used for the second and other axes, extra steps are involved. On the other hand, this computational scheme shows very clearly why the method is called Reciprocal Averaging.

MATRIX ALGEBRA APPROACH

For those with the necessary mathematical knowledge, the matrix algebra formulation of Reciprocal Averaging is given Appendix 1 of Hill's (1973) paper. As with PCA, RA produces its results in terms of eigenvalues and eigenvectors. The eigenvalues still decrease in value with successive axes, but the sum of the eigenvalues is not the same as the sum of the variances of either the species or the stands; consequently, the eigenvalues of RA are less important than they are in PCA. The elements of the eigenvectors are still, however, the axis scores required for the ordination diagrams.

Reciprocal Averaging is very similar to a non-centred, standardized by species, PCA. The real first axis, which does not appear in the scheme of elementary calculations described above, in fact corresponds to the first axis of a non-centred PCA, and its eigenvalue is always unity (the maximum). When a RA ordination is carried through by matrix algebra, the true first axis does appear with its eigenvalue of unity. Hence, the usual 'first' axis of RA is equivalent in this sense to the second axis of a non-centred PCA.

When RA was first introduced, Hill (1973) pointed out that its results were often very similar to those of a PCA on standardized data, and in data sets where this similarity did not exist, RA gave superior results from an ecological viewpoint than did PCA. However, the particular strength of RA was claimed to be the simultaneous ordination of both stands and species; but, as we have seen, PCA does this also. Subsequently, in comparative trials to be described at the end of this chapter, RA was shown to be superior to PCA for a number, though not all, of different data type situations.

Detrended Correspondence Analysis (Hill & Gauch 1980)

SHORTCOMINGS OF RECIPROCAL AVERAGING

Although an improvement on PCA for most data sets, RA still suffers from two major disadvantages. The first of these is the so-called 'arch effect', shown very well for the RA ordination of the Artificial Data in Figure 7.24a. This arch effect often appears in PCA ordinations also, but it seems to be particularly prevalent and pronounced in RA results. The important feature of this arch is that it is not usually a reflection of any ecological attribute of the data, and hence the vegetation structure, but is an artifact of the mathematical nature of the methods. It is, perhaps, not quite accurate to say that the arch effect has nothing to do with the data structure, because if the data had the statistical distribution of multivariate normality, no artificial arch effect would be obtained. However, as we know, scarcely any vegetation data sets are even remotely akin to a multivariate normal distribution, particularly if the data are qualitative. While it is true that the axes within any one PCA or RA ordination are orthogonal (no correlations between the axes), they are certainly not independent of one another; it is this lack of independence which, with typical vegetation data matrices (large excess of zeros), generates the arch effect. The non-independence of axes is not confined to the second (dependent on the first), but to all subsequent axes; and since an axis depends on all previous ones, the situation gets very complicated.

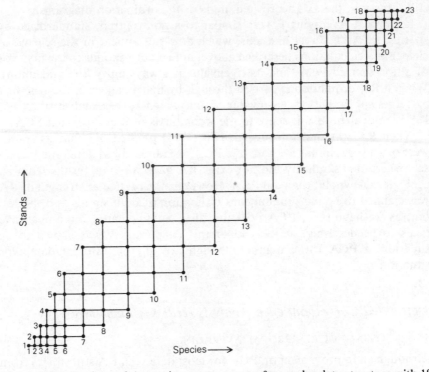

Figure 7.25 Reciprocal Averaging arrangement of a regular data structure with 18 stands in rows, spaced according to the first axis of a Reciprocal Averaging Ordination and, likewise, 23 species in columns. A species presence in a stand is indicated by a dot. Ideally, these stands would be spaced evenly (and likewise the species), but Reciprocal Averaging Ordination compresses the ends of the gradient relative to the middle. (After Hill & Gauch 1980, by permission of *Vegetatio*.)

The second undesirable feature of RA is shown in Figures 7.25 and 7.26. Figure 7.25 shows the RA arrangement on the first axis – of species in the horizontal direction, and of stands in the vertical direction – of an artificial regular data structure comprising 18 stands and 23 species. Ideally, all points should be equidistant in both directions, but RA compresses the points at the ends in both directions. This further means that the ends of the vegetational gradient represented by the first RA axis are compressed. Figure 7.26 shows both arch and compression effects in relation to the stand ordination only, for the same regular data matrix: Figure 7.26a shows the arch effect, involving the second axis, and Figure 7.26b demonstrates the compression effect of a single gradient (specifically Axis 1).

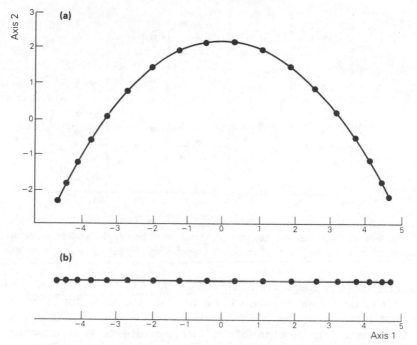

Figure 7.26 Two major faults of Reciprocal Averaging Ordination: (a) the arch distortion of the second axis; (b) compression of the ends of the first axis relative to the middle. These results are based on the data shown in Figure 7.25. (After Hill & Gauch 1980, by permission of *Vegetatio.*)

PRINCIPLES OF THE METHOD OF DETRENDED CORRESPONDENCE ANALYSIS

The method now to be described, Detrended Correspondence Analysis, has been designed to eliminate both the undesirable arch and compression effects. The method is essentially Reciprocal Averaging with two refinements, and we need only describe the latter in bare outline. The full technical details are given by Hill (1979a) and Hill & Gauch (1980).

Figure 7.27 illustrates the method of detrending the second axis. The first axis is divided into segments, and for each segment the mean of the second axis scores of the points therein is calculated. Again for each segment, the second axis stand scores are subtracted from their own segment mean, thus resulting in all second axis scores dispersed about a mean of zero. This detrending is applied to the sample scores at each iteration, except that, once convergence is reached, the final stand scores are derived by weighted averages of the species scores without detrending. This procedure results in a DCA ordination of the species with no arch problem, and a corresponding set of stand scores, which are simply weighted averages of the species scores

209

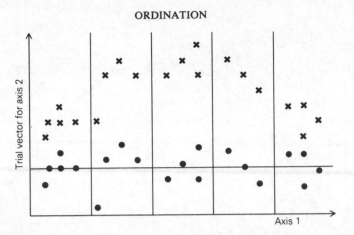

Figure 7.27 The method of detrending used in Detrended Correspondence Ordination. For explanation, see text. A stand score before detrending is shown as X, and after detrending as • (after Hill & Gauch 1980, by permission of *Vegetatio*).

(as in RA). To calculate a third DCA axis, stand scores are detrended with respect to the second axis as well as the first, and so on for higher axes (Gauch 1982).

Overcoming the problem of axis lengths distortions is achieved in principle by expanding (at the ends) and contracting (in the middle) small segments along the species ordination axes such that species turnover occurs at a uniform rate along the species ordination axes and, consequently, that equal distances on the stand ordination axes correspond to equal differences in species composition.

The basis for expansion and contraction is an attempt to equalize the mean within-stand dispersion of the species scores at all points along the gradient (axis). The relationship between dispersion and position along the gradient is shown in Figure 7.28; where the dispersion is high the gradient is contracted, and vice versa. Note that it is the *species* ordination axes that are adjusted, not the stand ordination, but the latter axes are defined at all times so that the stand scores are weighted mean values of the scores of the species that occur in them (Hill & Gauch 1980). Further details are given by Hill (1979a) and by Hill & Gauch (1980).

By means of the program DECORANA, the results of DCA of the quantitative Artificial Data are given in Table 7.16, and the first two axes (again standardized from 0–100) are plotted in Figure 7.29. These graphs should be compared with those of RA ordination of the same data (Fig. 7.24). It will be seen, for both stands and species, that the ordering along the first axis is the same in DCA as for RA although the relative distances between stands and between species differs, and also that there is no systematic variation on the second axis in relation to the first in DCA as

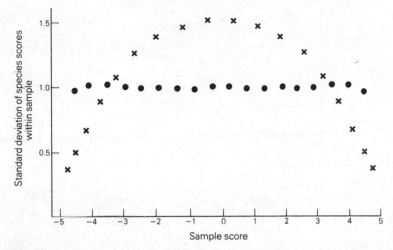

Figure 7.28 Within-stand standard deviation of species scores in relation to position along the first (stand) axis in Detrended Correspondence Ordination (●), contrasted with Reciprocal Averaging (X) (after Hill & Gauch 1980, by permission of *Vegetatio*). For explanation, see text.

Table 7.16 The first three axes of the Detrended Correspondence Ordination of the quantitative Artificial Data.

	Axis 1	Axis 2	Axis 3
eigenvalue	0.7896	0.0721	0.0245
stand			
1	31	0	93
2	156	62	0
3	0	62	100
4	196	14	68
5	194	57	23
6	132	33	30
7	273	9	112
8	4	33	76
9	142	9	46
10	98	60	21
species			
I	112	− 101	144
II	161	78	− 14
III	4	33	76
IV	273	9	112
V	− 304	227	− 160

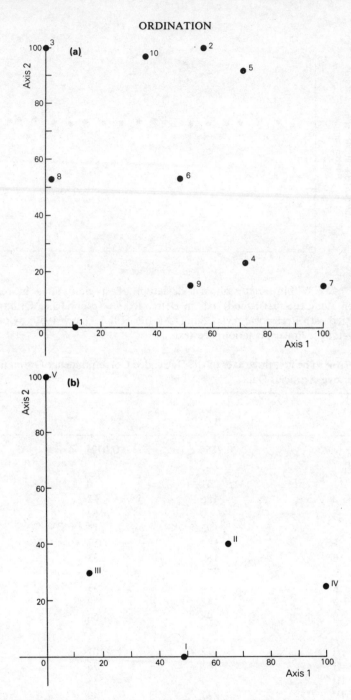

Figure 7.29 Detrended Correspondence Ordination of the quantitative Artificial Data: (a) stands; (b) species.

there is in RA. Certainly the arch effect has been eliminated, and presumably any artificial compressions of the axes ends have disappeared too.

Comparisons of ordination methods

Ordination methods are expected to arrange stands (or species) in relation to graphical axes or, equivalently, vegetation gradients. The axes may often reflect environmental gradients, and an important purpose of ordination is to try and identify the axes in environmental terms. This purpose is defeated if there are distortions in one or more of the axes produced by an interaction of the mathematics of the method with the data structure. The distortion does, in fact, relate to the ecological properties of the stands in the data, and these properties and the nature of the distortions produced by ordinations will be discussed below.

Real and simulated data

In order to compare ordination methods from the viewpoint of their ecological effectiveness, they must be tried on different data sets. Using sets of real field data for this purpose is, however, not entirely satisfactory. This is because we judge an ordination to be good if we can interpret it ecologically; but to do this requires a subjective judgement of the ecological background of the habitat surveyed. This is not to say that sensible comparisons of ordination methods cannot be made with field data; indeed, this will be usefully done for the Coed Nant Lolwyn data in Chapter 9: but ordination methods are essentially applied to the complex situation of the field, and in this sense, field data have the ideal property – realism. But in field data we do not *know* what the underlying structure is, otherwise we should not be analyzing the data in the first place; we can only surmise what the structure might be from the current ordination and previous knowledge.

There is another kind of data – artificial or **simulated data.** Such data can have any structure we please, including a random element built in if we so desire.

The basis of artificial data simulating a field situation is the **coenocline,** which is defined as a gradient in community composition (Gauch 1982). An ideal realization of this definition is shown in Figure 7.30, in which are shown curves of species abundance along an environmental gradient with sample stands taken at even intervals along the gradient. These curves are actually normal distributions, and they are often known as **Gaussian curves.**

Coenoclines may vary enormously in their *species turnover* from one end to another. Species turnover can be loosely described as species 'coming in and out' as one moves along the coenocline, as can be appreciated from

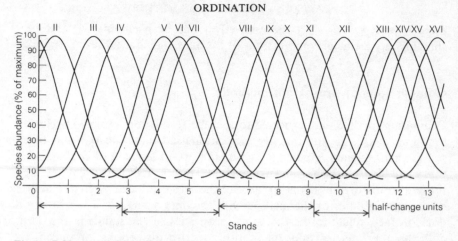

Figure 7.30 A simulated coenocline with 17 species and 13 sample stands. Single half-changes are shown along the coenocline, calculated on the basis of Czekanowski's similarity coefficient.

examination of Figure 7.30. The amount of species turnover in a coenocline is also known as **beta diversity.** One way of measuring beta diversity in a coenocline is by specifying the number of half-changes occurring from one end to the other. One **half-change** is defined as the separation along the coenocline at which the similarity between two stands is 50%. The number of half-changes from the left-hand end of the coenocline is shown in Figure 7.30.

Similar ideas can be extended to two distinct gradients, which may be depicted at right angles to one another as a **coenoplane.** Now the Gaussian curves become Gaussian surfaces (rounded cones) placed on the coenoplane in two directions. Half-changes can be specified in each direction; so we can have a square coenoplane where the beta diversity is about the same for the two gradients, or a rectangular coenoplane in which the beta diversity differs in each direction.

A comparison of ordination methods based on simulated data

A very revealing comparison of the performance of several ordination methods, based on simulated data, was made by Gauch *et al.* (1977). It is worth describing their work in detail not only for the insight provided on the efficacy of various ordination methods but also for the assistance the study gives in the interpretation of ordinations of real field data.

The quantitative simulated data used comprised coenoclines containing differing numbers of half-changes, and coenoplanes which varied in shape from square to thinly rectangular. The main ordination methods compared

214

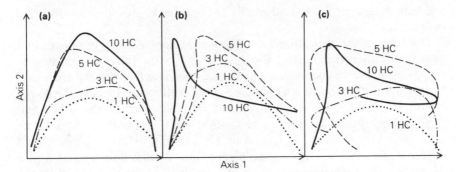

Figure 7.31 Distortions of simulated coenoclines into arches by: (a) Reciprocal Averaging; (b) centred and species-standardized PCO; (c) centred and non-standardized PCO. Sample ordinations for four levels of beta diversity (1, 3, 5, 10 half-changes) are shown; and in all cases the result should be a horizontal straight line. (After Gauch *et al.* 1977, by permission of the *Journal of Ecology*.)

were: Reciprocal Averaging (RA); species-centred Principal Component Ordination (PCO) in both non-standardized and species-standardized forms; and Bray & Curtis Ordination (BC). Orloci Ordination was occasionally tried. The method of Detrended Correspondence Ordination had not yet been formulated but, since this method was devised to overcome the deficiencies of RA, the work of Gauch *et al.* loses none of its value by the absence of this technique.

COENOCLINE RESULTS

Twelve coenoclines were simulated, ranging in beta diversity from 0.3 to 20 half-changes. Each coenocline had about 20 simulated species and 24 sample stands placed uniformly along the gradient. The ideal ordination should recover this structure in the form of evenly spaced sample points along the first axis, with no differential scores on the second or higher axes.

Figure 7.31 shows the results obtained for RA, standardized PCO, and non-standardized PCO. In all cases there is distortion, both in the form of an arch into the second axis, and in the varying differences between Axis 1 scores of stands at the ends of the coenocline (or Axis 1) as compared with stands in the middle. These are features of an ordination which have been described previously (p. 207). The degree of distortion increases with increasing beta diversity.

The arch effect seems to be slightly worse in standardized PCO than in RA. In non-standardized PCO not only is there an arch effect, but at three half-changes and greater there is *involution* of one or more of the axis ends. This means that the stands at the extreme ends of the coenocline are not so placed on the first ordination axis − an additional distortion. However for

215

one half-change, non-standardized PCO appeared to give the least distorted ordination.

Even though the coenocline has only one gradient, distortion extends into axes higher than the second. Figure 7.32 shows the relationship between stand positions on a coenocline and axis scores on each of the first eight axes for the RA ordination of a five half-changes coenocline. Ideally, the Axis 1 relationship should be a diagonal straight line, and all the other axis relationships should be horizontal straight lines. Apart from Axis 1, the actuality is very different. In fact, Axis n has an approximate polynomial of degree n relationship with coenocline position. As axis number and degree of polynomial both increase, less distance on the axis is traversed by the distortion; but it is not until the eighth axis is reached that the distortion becomes negligible for this set of data.

Another way of viewing the distortions is shown in Figure 7.33a–c. This shows, again for a coenocline of five half-changes, a non-standardized PCO involving all combinations of the first three axes. The ideal ordination would give horizontal straight lines in Figures 7.33a and b, and simply a point on the plane formed by Axes 2 and 3 in Figure 7.33c.

Although not explicitly shown in their paper, Gauch et al. imply that Bray & Curtis Ordination is not usually subject to any of the distortions discussed above. First, because the axes are defined by stands with maximum dissimilarity, i.e. at opposite ends of the coenocline, involution is

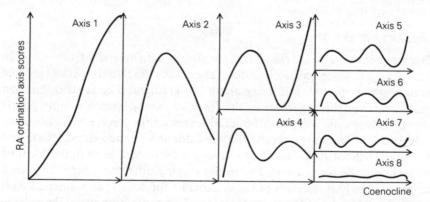

Figure 7.32 Reciprocal Averaging Ordination of a 5 half-changes coenocline to eight dimensions (after Gauch et al. 1977, by permission of the Journal of Ecology).

Figure 7.33 (a)–(c) centred and non-standardized PCO of a 5 half-changes coenocline in three dimensions, the 24 stands being evenly spaced along the coenocline; (d)–(f) the same ordination for a 2.5×1.5 half-changes coenoplane, the 40 stands being arranged in an evenly spaced grid (after Gauch et al. 1977, by permission of the Journal of Ecology).

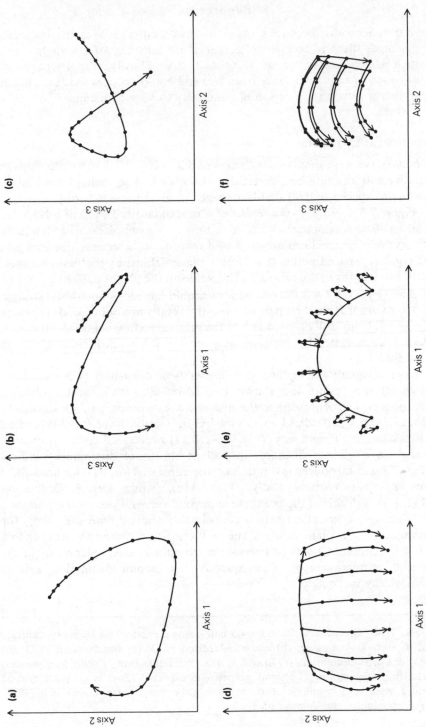

avoided. Secondly, because stand positions on different axes are separately calculated, there is no coenocline curvature into any of the higher axes which are characteristic of PCO and RA. Thirdly, the Pythagorean projection of sample distance from two end points reduces the effect of the curvilinear relationship of sample similarity to sample distance.

COENOPLANE RESULTS

Coenoplanes were simulated with axes of 0.7 to 10 half-changes, varying the relative and absolute beta diversities of the axes. The coenoplanes had 30 simulated species and 40 stands arranged in a regular 8×5 grid.

Figure 7.33d–f shows the results of non-standardized PCO of a 2.5×1.5 half-changes coenoplane – i.e. a rectangular coenoplane with low beta diversities. The ideal ordination would reproduce the coenoplane as a grid of regularly spaced points (8×5) on a plane defined by the first two axes, with axis lengths in the ratio $2.5 : 1.5$; while on the planes defined by Axes 1 & 3 and by Axes 2 & 3 the ordinations should appear as horizontal straight lines. As regards the first pair of axes, the results are quite good; the main distortion being a slight rounding of the original rectangular grid. However, Axis 3 is dominated by the spurious arch. The RA result is similar, though less distorted.

For a square coenoplane with higher beta diversities (4.5×4.5 half-changes) the results are shown in Figure 7.34. Reciprocal Averaging reproduces the coenoplane in the first two axes with only minor distortion (Fig. 7.34a); standardized PCO and BC ordination have small distorting effects, but in different ways (Fig. 7.34c,d). However, using BC Ordination based on Euclidean distances instead of the usual dissimilarity indices (Fig. 7.34e), introduces involution of the corners. Non-standardized PCO involutes the corners badly (Fig. 7.34b), while Orloci Ordination (Fig. 7.34f) distorted the coenoplane beyond recognition.

When one coenoplane axis is considerably shorter than the other, for example 4.5×1.5 half-changes, the arch distortion caused by the long axis of the coenoplane tends to overwhelm the shorter coenoplane axis in the second ordination axis. Consequently the second ordination axis is ecologically meaningless.

Figure 7.34 Ordination of a 4.5×4.5 half-changes coenoplane by six techniques: (a) Reciprocal Averaging; (b) non-standardized PCO; (c) standardized PCO; (d) Bray & Curtis Ordination; (e) Bray & Curtis Ordination using Euclidean distances; (f) Orloci Ordination. The stand pattern and expected result is a square grid of points, eight rows in one direction, and five in the other. (After Gauch *et al.* 1977, by permission of the *Journal of Ecology*.)

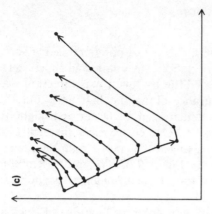

(a)

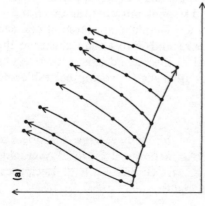

(b)

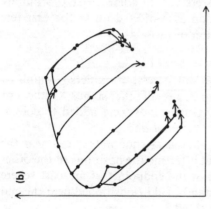

(c)

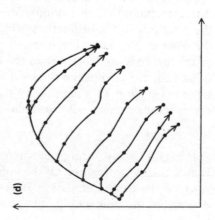

(d)

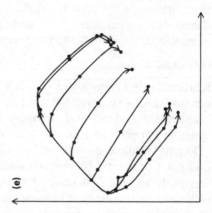

(e)

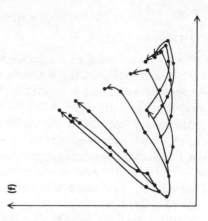

(f)

CLUSTERS OF STANDS

Some coenoplane data were created with clusters of points replacing some of the single points. Bray & Curtis Ordination was unaffected by clusters. Reciprocal Averaging was affected very little by them; axis direction was rotated slightly to align maximal variance along the first axis, but the configuration was fairly rigid. Non-standardized PCO showed slight to extensive deterioration with stand clusters. One cluster near the centre of the 2.5×1.5 half-changes coenoplane changed the Axes 1 and 2 configurations drastically, from one resembling Figure 7.33d to one resembling Figure 7.33e. The same test with the 4.5×1.5 half-changes coenoplane showed little effect.

Clusters tend to attract the axes of non-polar ordinations. Principal Component Ordination is more vulnerable than RA to this effect. Square coenoplanes seem to be most vulnerable to the distorting effect of clusters; axis rotations produced by stand clusters can in some cases profoundly change the appearance of the ordination as projected on to the first few axes, and consequently the interpretation of those axes.

OUTLIERS

Real data often include one to a few deviant stands, or **outliers** – stands of unusual species composition when compared to all other stands. Outliers of 'moderately deviant' and 'strongly deviant' stands were added to coenoclines and coenoplanes in the work of Gauch *et al*.

For BC Ordination, all outliers not chosen as endpoints occur near the mid-point of the ordination axis without affecting the positions of the other stands. If an outlier is selected as one of the endpoints of an axis, severe distortion occurs because most of the stands will be positioned near the end of the axis defined by the non-deviant stand.

Reciprocal Averaging was robust for moderately deviant outliers – they appeared in the centre of the ordination field and caused negligible displacement of the other stands. Strongly deviant outliers appeared around the periphery of the ordination field and strongly affected the positions of the other stands. One such outlier from a coenoplane appeared at one end of the first axis and compressed the other stands into a tight cluster at the other end; the coenocline then emerged in the second axis. Other results given in Gauch *et al*.'s paper show that outliers can have unpredictable and severe results on non-polar ordinations.

DISCUSSION

Perhaps the biggest surprise to emerge from this study was the robustness of Bray & Curtis Ordination. Of the non-polar methods, Reciprocal Averaging emerged the best. Comparing Bray & Curtis Ordination with Reciprocal Averaging, the following points can be made.

(a) BC Ordination is immune from the arch effect, whereas RA is not.
(b) RA is immune from involution; BC Ordination is normally immune from involution provided care is taken in choosing the end-point stands to define the axes.
(c) BC Ordination reproduces true inter-stand distances along a coenocline rather better than does RA.
(d) BC Ordination is unaffected by clusters of stands, whereas RA could be slightly affected.
(e) Both RA and BC Ordinations can be equally severely affected by outliers.

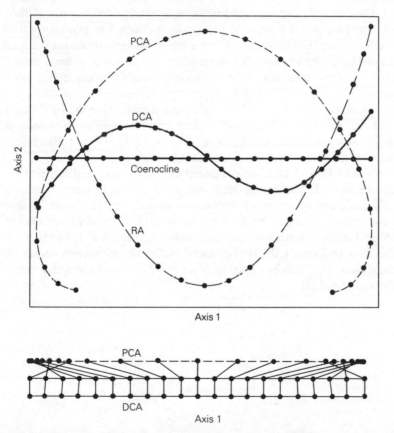

Figure 7.35 Representation of a simulated coenocline by three ordination techniques: non-standardized PCO; Reciprocal Averaging; Detrended Correspondence Ordination. The coenocline has 21 species and 21 evenly spaced stands, and with a gradient length of 5 half-changes. The lower panel shows the ordinations on Axis 1 only, while the upper panel shows Axes 1 and 2. (After Gauch 1982, by permission of Cambridge University Press.)

Thus it would seem that BC Ordination is, on the whole, superior to RA; but not in every situation. The main disadvantage of BC Ordination is that it is not a computationally automatic procedure. This is no great disadvantage for small data sets with all the interactive computing facilities now available, but it makes the ordination of large data sets very much more difficult as there is usually considerable choice of stand pairs which could be used to define the ends of each axis. However, for small data sets, Bray & Curtis Ordination is now seen as an attractive and effective ordination method.

Among the non-polar methods there would appear to be a decline in efficiency through the series: RA, standardized PCO, non-standardized PCO. Only in the case of very low beta diversity could non-standardized PCO be recommended, since in this study it seemed to produce less of an arch effect than the other two. For medium-low beta diversities, standardized PCO could be a reasonable alternative to RA, but in general there is no doubt that of the three non-polar methods compared in this study, RA is the best.

The method of Detrended Correspondence Ordination has been formulated to eliminate the arch effect and the end-of-axes inter-stand distance distortions shown by RA. The achievement of these aims is admirably demonstrated for a simulated coenocline in Figure 7.35. Presumably, however, difficulties produced by clusters and outliers remain. Of these two possible features of field data, outliers would seem to present the greater problem in terms of ordination distortion. However, a severely deviant outlier can easily be detected on an ordination diagram since it lies at one end of an axis with the remaining stands clustered at the other end. It is thus easy to identify the outlier stand and study its species composition. Then, if desired, it can be removed from the data matrix and the remaining stands reordinated.

8

Correlations between vegetation and environment

In this chapter we examine methods suitable for correlating vegetation features with environmental factors in stands or groups of stands. There are a variety of situations which need to be considered; and we shall proceed in increasing order of complexity, from the situation of qualitative records of a single species with qualitative records of an environmental factor in single stands, to quantitative records of many species with quantitative records of several environmental factors in whole associations.

Methods of examining correlations may produce results in two different forms: graphical, or a number indicating the degree of correlation which may or may not be suitable for a statistical test. Remember that, strictly, for statistical tests to be valid the sampling scheme should be random; but see the discussion on page 17.

Single species

Single environmental factor

QUALITATIVE SPECIES AND QUALITATIVE ENVIRONMENTAL RECORDS

A single species and a single environmental factor would rarely comprise an entire data set; much more commonly such data are subsets of a more comprehensive data set, and the example below is abstracted from a complete vegetation survey of an area of chalk grassland. The method of assessing correlation is to use a 2×2 contingency table where the presence and absence of the species constitutes one variate, and the presence and absence (or any other two states) of the environmental factor is the other variate.

223

Example 8.1

In chalk grassland, the species *Asperula cynanchica* (Squinancy Wort) is said to occur preferentially on ant-hills. The following data are extracted from a survey in which the stands of size 0.5×0.5 m were selected every 2 m in each direction on a 30×6 m grid. Although the stands were sampled in a regular fashion, if the ant-hills occurred on the ground at random then the samples would be random with respect to the presence or absence of ant-hills. Even if the ant-hills did not have a random spatial distribution, the method of sampling did not take into account ant-hill distribution, and so the sampling is effectively random with respect to the occurrences of ant-hills in the stands. The data, expressed in contingency table form, are as follows:

		Asperula cynanchica		
		+	−	
Ant-hill wholly or partially within stand	+	2	5	7
	−	11	54	65
		13	59	72

and

$$\chi^2_{[1]} = \frac{72(53)^2}{(13)(59)(65)(7)} = 0.58$$

The result is non-significant, and so there is no evidence of any association; the corresponding correlation coefficient is given by

$$(\chi^2/n) = (0.58/72) = 0.09$$

The method could be extended to the situation where there are more than two states of the environmental factor or where the quantity of a species is divided into a number of groups. For a states of the environmental factor, and b groups of species abundance, we would have an $a \times b$ contingency table. The interpretation of the result would not, however, be as straightforward as for the 2×2 case.

QUALITATIVE SPECIES AND QUANTITATIVE ENVIRONMENTAL RECORDS

In this very common category, the species records are presence or absence, but the environmental factors are those which can be measured on a continuous scale.

Graphical presentation   A graphical presentation is, in effect, a direct gradient analysis.

Example 8.2

The histogram in Figure 8.1 shows the relationship between the frequency of occurrence of *Anemone nemorosa* in soils of different extractable phosphorus contents in the Coed Nant Lolwyn case study. The reasons for having different phosphorus content ranges have been given in Chapter 7 (p. 156). In this data set, *Anemone nemorosa* is the commonest species, with relative frequency 79%, but there is a clear decrease in its relative frequency of occurrence at higher soil phosphate levels. Compare this presentation with that of Figure 7.6b.

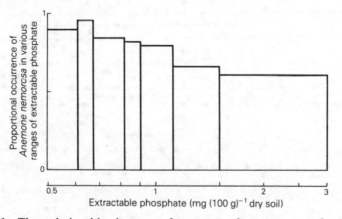

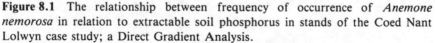

Figure 8.1 The relationship between frequency of occurrence of *Anemone nemorosa* in relation to extractable soil phosphorus in stands of the Coed Nant Lolwyn case study; a Direct Gradient Analysis.

Example 8.3

The occurrence of a single species in relation to two environmental factor gradients can be shown on one graph. Figure 8.2 again shows the distribution of *Anemone nemorosa* in the Coed Nant Lolwyn data, this time in relation to both soil phosphate (P) and vernal light intensity (L2). The area of the graph has been divided into four quadrants, with the dividing lines established at the means (geometric means, because of the use of the

225

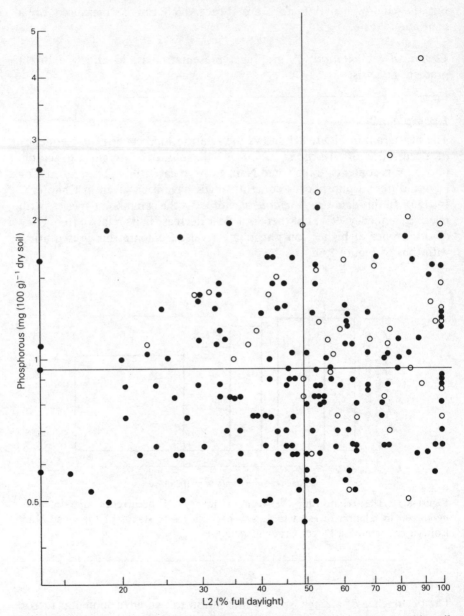

Figure 8.2 The occurrence of *Anemone nemorosa* in relation to extractable soil phosphorus and vernal light intensity in the Coed Nant Lolwyn case study; a Direct Gradient Analysis *sensu stricto*. Stands are shown as circles: *Anemone nemorosa* present (•), *Anemone nemorosa* absent (○).

environmental data in logarithmic form) of phosphate and vernal light intensity, respectively, and the relative frequency of occurrence of *Anemone nemorosa* evaluated in each of the four quadrants. The percentage frequencies of occurrence of *Anemone nemorosa* in the four quadrants of the graph are as follows: lower left (low P, low L2), 96; lower right (low P, high L2), 81; upper left (high P, low L2), 77; upper right (high P, high L2), 60. Evidently, this species is favoured by both low soil phosphorus and low vernal light levels. This example is, in effect, a form of two factor direct gradient analysis *sensu stricto*.

Comparison of a species' frequency in stands having levels of an environmental factor greater and less than the mean   This simple method enables one to make a statistical test of the propensity of a species to occur at higher or lower levels of the measured range of an environmental factor. Divide the total number of stands into two groups: one group having stands whose levels of the environmental factor in question are greater than the mean level of that factor across all the stands, and the other group whose stands have lower levels than the mean. Then construct a 2×2 contingency table:

		Environmental factor		
		> mean	< mean	
Species	present	a	c	$a + c$
	absent	b	d	$b + d$
		$a + b$	$c + d$	$a + b + c + d$

and analyze it in the usual way (p. 77–8).

Example 8.4

In the Coed Nant Lolwyn data, *Ranunculus ficaria* occurs in 141 out of the 200 stands. In relation to soil pH, whose mean value for all stands is 4.85, 96 stands have a soil pH greater than the mean and 104 stands have levels of soil pH less than the mean. Out of the 96 stands of higher than average soil pH, 74 contain *Ranunculus ficaria*; and of the remaining 104 stands, 67 contain this species. The contingency table is:

		Soil pH		
		>4.85	<4.85	
Ranunculus *ficaria*	+	74	67	141
	−	22	37	59
		96	104	200

giving

$$\chi^2_{[1]} = \frac{200(1264)^2}{(96)(104)(141)(59)} = 3.85^*$$

which is just significant at P(0.05). Hence there are significantly more occurrences of *Ranunculus ficaria* in stands of soil pH greater than 4.85 than in stands of soil pH less than this value.

At first sight it might be thought that the correct way to evaluate the chi-square would be to simply compare the observed number of stands of greater than average level of the environmental factor with the expected number which contain the species, and similarly for the stands of lower than average level of the factor concerned. For the situation in Example 8.4, the relative frequency of *Ranunculus ficaria* in all 200 stands is 0.705. Hence the expected number of stands of greater than average soil pH to contain this species, assuming no correlation of occurrence with soil pH, is given by $96 \times 0.705 = 67.68$; and, similarly, the number of stands of soil pH less than 4.85 expected to contain *Ranunculus ficaria* is given by $104 \times 0.705 = 73.32$. Then

$$\chi^2_{[1]} = \frac{(74 - 67.68)^2}{67.68} + \frac{(67 - 73.32)^2}{73.32} = 1.14$$

which is nowhere near significant.

That this argument is wrong can be appreciated by considering an example which is the converse of the above. Suppose, out of the 200 stands, a species occurred in 59 of them giving a relative frequency of 0.295. Let us further suppose that 22 of the 59 occurrences were in the 96 stands of greater than average soil pH, and 37 occurrences of the species were in the 104 stands of less than average soil pH. The expected number of occurrences in the first group of stands is $96 \times 0.295 = 28.32$, and in the second group of stands is $104 \times 0.295 = 30.68$; then

$$\chi^2_{[1]} = \frac{(22 - 28.32)^2}{28.32} + \frac{(37 - 30.68)^2}{30.68} = 2.71$$

Although still not significant, χ^2 is more than twice as large.

The situation above is exactly the converse of that in Example 8.4: what are presences in one case are absences in the other, but the final result *should* be the same in both instances. To achieve this, we have to consider not only those stands which contain *Ranunculus ficaria* (or the hypothetical species) but also those stands not containing the species. However, it would be wrong to write

$$\chi^2_{[3]} = \frac{(74 - 67.68)^2}{67.68} + \frac{(67 - 73.32)^2}{73.32} + \frac{(22 - 28.32)^2}{28.32} + \frac{(37 - 30.68)^2}{30.68} = 3.85$$

i.e. chi-square with 3 degrees of freedom, since the four terms in the χ^2 summation are dependent on one another. This is because altering only one observed value out of the four would also change the other three. Hence there is only 1 degree of freedom, and this is why the contingency table method is used.

The final point to note is that the level of environmental factor which partitions the stands into two groups does not have to be the mean; it can be any specified level.

Comparison of the mean levels of an environmental factor in stands containing and not containing a particular species In one respect, this test is akin to the last in that the stands are divided into two groups; but this time the groups comprise on the one hand those stands containing the species in question, and on the other those stands which do not. The mean levels of the relevant environmental factor are then compared by means of a two-sample Student's t-test.

Example 8.5

Again, we shall use the species *Ranunculus ficaria* in the Coed Nant Lolwyn data in relation to soil pH. The relevant details for the stands containing the species are:

$$n_1 = 141 \quad \Sigma x_1 = 694.40 \quad \Sigma x_1^2 = 3442.0822 \quad \bar{x}_1 = 4.93 \quad s_1 = 0.3990$$

and for stands not containing *Ranunculus ficaria*, the corresponding quantities are:

$$n_2 = 59 \quad \Sigma x_2 = 257.07 \quad \Sigma x_2^2 = 1297.0103 \quad \bar{x}_2 = 4.66 \quad s_2 = 0.5013$$

We should first assess whether the variances of the two samples differ significantly: we have that $s_1^2 = 0.1592$ and $s_2^2 = 0.2513$, hence

$$F_{[58,140]} = \frac{0.2513}{0.1592} = 1.58^*$$

which is just significant at $P(0.05)$. Remember that the 2.5% F-table has to be used for a probability of 5%, because this use of the F-ratio is a two-tail test; but the probability levels of F-tables are specified assuming a one-tail test, which is appropriate for their usual employment in the analysis of variance.

Since the two samples cannot be presumed to come from underlying populations of soil pH-values having identical variances, the following approximate form of t-test must be used:

$$t_{[198]} = \frac{4.93 - 4.66}{(0.1592/141 + 0.2513/59)} = 3.68^{***}$$

which is significant at $P(0.001)$, thus again showing that *Ranunculus ficaria* tends to grow on soils of higher pH.

QUANTITATIVE SPECIES AND QUANTITATIVE ENVIRONMENTAL RECORDS

The problems here are similar to those encountered in calculating associations between species for quantitative data (Ch. 5, p. 89–90). The product-moment correlation coefficient is a useful index of correlation, but is not suitable as a significance test of correlation because the data could scarcely approximate to a bivariate normal distribution. A rank correlation coefficient is much superior in this situation, and the following example yet again uses the Coed Nant Lolwyn *Ranunculus ficaria* and soil pH data.

Example 8.6

The data consist of 200 pairs of values, each pair consisting of a soil pH level and a density m^{-2} of *Ranunculus ficaria*. Of the latter, 141 are non-zero (corresponding to presences), and the remaining 59 are zeros (corresponding to absences). The working is too voluminous to reproduce here, and in any case too laborious to compute by hand; so we merely quote the results, which were obtained from a computer program. Spearman's correlation coefficient is $r_S = 0.05$ with a corresponding Student's t-value of

$$t_{[198]} = 0.75$$

Evidently there is no correlation between soil pH and density of *Ranunculus ficaria* although, as we have seen, two different ways of treating the qualitative data shows there to be a significant positive association between the *presence* of *Ranunculus ficaria* and higher soil pH-values.

Use of the species density data in the above example shows a disadvantageous feature of quantitative data. The clear association of the occurrence of *Ranunculus ficaria* with soils of higher pH in Coed Nant Lolwyn has been masked by using the density data. This does not mean that the quantitative data should not be examined, but that it should be used in conjunction with the qualitative data in order to obtain a full and clear picture of the nature of an association between a species and an environmental factor. For the current situation, Examples 8.4 and 8.5 show a clear positive association between the presence of *Ranunculus ficaria* and soils of higher pH; but the abundances within stands, in terms of plant density, of the species appears to be independent of this environmental factor.

Several environmental factors

HOTELLING'S T^2

Hotelling's T^2 test is a generalization of Student's t-test in which, as before, the whole group of stands is divided into two sub-groups – those stands which do, and those that do not contain the species in question. Then, instead of comparing the two means of a single environmental factor, two column vectors, each containing the means of several environmental factors, are compared as a whole. Although the technique is simple in theory, there may be computational difficulties because the method requires extracting the eigenvalues and eigenvectors of a non-symmetric matrix. Even with a computer this may present a numerical problem, and until this problem has been overcome, the use of Hotelling's T^2 cannot be regarded as a practical proposition. The best way at present to treat several environmental factors is to deal with them one at a time by the methods of the previous section.

Several species

A single environmental factor

In the rare cases when only one environmental factor has been assessed in a vegetation survey, it is best to treat one species at a time by the methods of the previous part of the chapter; or some of the appropriate methods of the following sections can be tried.

Several environmental factors

CORRELATION WITH ORDINATION AXES

As previously explained (Ch. 7, p. 146), at least some of the axes of ordinations of vegetation data can be presumed to reflect environmental gradients. Even if no environmental data are available from the stands in a vegetation survey, attempts are usually made to interpret some of the lowest axes environmentally by noting the positions (scores) of species whose ecological attributes are partially known. Then hypotheses are made about other species according to their scores on the axes which have been given tentative environmental meanings. However, when the purpose of a survey is to investigate species–environment relationships, such a procedure becomes a circular argument: environmental data are now essential.

Armed with stand environmental data, it may be possible to environmentally characterize one or more of the ordination axes. Since each stand has both a level of each environmental factor measured and a score on each ordination axis, correlation coefficients may be calculated for each environmental factor level and ordination axis score pairs. If a nonparametric rank correlation coefficient is used, it can also be statistically assessed.

Example 8.7

Figure 8.3 shows a Detrended Correspondence species ordination of the Coed Nant Lolwyn data. Only those species which have a relative frequency of occurrence of at least 10% have been used for the ordination (Minimum Number of Occurrences (MINO = 20) in order to present a relatively simple situation.

From the stand ordination, Spearman's rank correlation coefficients were calculated between the environmental factors measured in the stands and the stand scores on the ordination axes. Table 8.1 shows the details of those coefficients which differ significantly from zero for Axes 1 and 2.

From the viewpoint of the *measured* environmental factors it would seem that Axis 1 is reflecting a gradient of base richness, with manganese levels

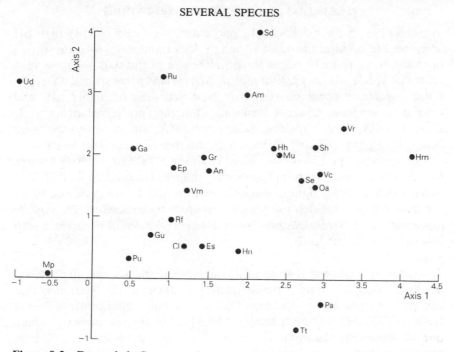

Figure 8.3 Detrended Correspondence species ordination of the Coed Nant Lolwyn case study. Only those species with a relative frequency of at least 10% have been included in the analysis.

Table 8.1 Spearman's rank correlation coefficients between environmental factor levels and Axes 1 and 2 scores of Detrended Correspondence Ordination of the Coed Nant Lolwyn data. Only the 27 most frequent species are employed in the ordination, and only correlations which are significantly different from zero are shown.

Axis 1		Axis 2	
pH	−0.283***	phosphorus	0.247***
calcium	−0.247***	manganese	0.170*
pre-vernal light	−0.179*	....	
sodium	−0.178*	pH	−0.149*
magnesium	−0.153*		
....			
phosphorus	0.167*		
manganese	0.211**		

233

in opposition to base richness. In particular, high calcium and high pH levels tend to occur at the negative end of Axis 1 and vice versa. Pre-vernal light also tends to be higher at the negative end of this axis. Thus we may infer that species such as *Urtica dioica*, *Mercurialis perennis* and *Plagiomnium undulatum* occur on the more base-rich soils of higher pH, and possibly a tendency towards locations of higher pre-vernal light; while *Holcus mollis*, *Viola riviniana*, *Oxalis acetosella*, etc. occur on the more acidic, base-poor soils and perhaps with lower pre-vernal light levels.

Again, with respect to the measured environmental factors, Axis 2 seems to represent essentially a soil phosphorus gradient, from which we infer that species such as *Silene dioica*, *Rubus fruticosus* and *Urtica dioica* occur on the more phosphorus-rich soils; with *Thuidium tamariscinum*, *Plagiochila asplenoides* and *Mercurialis perennis* having correlations of occurrence with low soil phosphorus levels.

The statements made at the end of the above example are only hypotheses, not facts. Although some of the correlation coefficients, i.e. those involving soil pH, calcium and phosphorus, are very highly significantly different from zero, they are still quite small ($< |0.3|$). The correlation between any one of these environmental factors and an ordination axis is not even remotely perfect.

Another point to bear in mind is that the axes of an ordination carried out on vegetation data are *vegetation gradients*, not environmental gradients; the axes only reflect such of the measured environmental gradients which are correlated with species occurrences, and only a very small sub-set of the total environment can be practically assessed in a study. This feature is well brought out by Axis 2 in the above example which seems, from a vegetation and stand viewpoint, to represent a wood edge to wood interior gradient. An indefinitely large number of environmental factors would contribute to this gradient, producing the vegetation changes characterizing this ordination axis, but the only one factor actually measured in this study which is relevant to this gradient is soil phosphorus, with small contributions from soil manganese and pH.

At first sight it seems strange that soil phosphorus is relevant to this gradient but that none of the light levels are; but this is probably due to the particular environmental characteristics of Coed Nant Lolwyn (Ch. 4). The long length of wood edge is a boundary with agricultural land at a higher elevation, and it may be that phosphorus from fertilizer applications has moved into the soils of Coed Nant Lolwyn, with progressively lower concentrations away from the wood edge. With regard to light intensity, the speculation has already been advanced that, owing to the narrowness and topography of the wood, light from the sides infiltrates most of the study area, so that positions near the wood edge are not conspicuously lighter (especially in spring) than those further in when considered over a whole 24-hour period.

SPECIES ORDINATION BASED ON ENVIRONMENTAL ATTRIBUTES

Example 8.7 gave rather tantalizing indications of species occurrences in relation to some environmental factors through correlations of environmental factors with axis scores of a vegetation data ordination. Evidently there are some relatively high correlations between the environmental factors of Coed Nant Lolwyn (see Table 8.2), and one way of utilizing these is to ordinate the environmental data directly. This is, in fact, the general method we have called 'Semi-direct Gradient Analysis', and PCA is a useful specific technique. We can still obtain stand and species ordinations, but now we are ordinating the environmental data themselves, and the species ordination involves a secondary set of computations which can be applied to a stand ordination obtained by any method.

Example 8.8

The same data, originating from Coed Nant Lolwyn, as were used in Example 8.7 are employed here. Table 8.2 shows the whole list of environmental factors together with their matrix of correlation coefficients, and Table 8.3 gives the results of a PCA carried out on the correlation matrix. This matrix has been used so that each environmental factor has the same weighting.

The first two components together account for 37.6% of the total variation. The first component is evidently highlighting a gradient of base richness (as reflected by the first axis of the vegetation ordination in Example 8.7), with the group of relatively high positive loadings of soil calcium, magnesium, sodium and pH. The second axis has high negative loadings of soil potassium, phosphorus and manganese. This axis could be a reflection of a soil texture gradient, with a tendency for stands with the more clayey soils to lie at the negative end of the axis.

Axis 3 represents a contrast between pre-vernal and vernal light levels on the one hand, and late aestival light and soil moisture levels on the other. Bearing in mind the difficulty associated with the circumstances under which the late aestival light measurements were made, it would be best to discount this factor and interpret Component 3 as a general spring light gradient. Axis 4, with its high loading of late aestival light, is likewise ignored. Axis 5, however, is essentially one of aestival light gradient, and with a weak contrast of pre-vernal light.

Figure 8.4 gives the corresponding species ordination, using the same sub-set of species as in Example 8.7. The position of a species in relation to a particular axis or, to put it another way, the score of a species on a

235

Table 8.2 List of the environmental factors, and their correlation matrix, in the Coed Nant Lolwyn data.

Number	
1	soil pH
2	soil phosphorus
3	soil potassium
4	soil magnesium
5	soil manganese	logarithmically transformed
6	soil calcium
7	soil sodium
8	soil organic matter (%)
9	soil moisture (%)
10	light, pre-vernal
11	light, vernal	logarithmically transformed
12	light, aestival
13	light, late aestival

Factor

	1	2	3	4	5	6	7	8	9	10	11	12	13
1	1.00												
2	−0.23	1.00											
3	−0.12	0.25	1.00										
4	0.46	0.12	0.44	1.00									
5	−0.46	0.21	0.27	−0.03	1.00								
6	0.64	0.01	0.07	0.63	−0.26	1.00							
7	0.47	0.04	0.15	0.63	−0.19	0.60	1.00						
8	−0.22	0.12	0.14	0.08	0.14	−0.04	−0.04	1.00					
9	−0.07	0.09	0.02	0.14	0.03	0.19	0.06	0.17	1.00				
10	0.14	−0.01	−0.05	0.03	−0.07	0.13	0.10	−0.09	−0.20	1.00			
11	−0.01	0.09	0.06	−0.05	0.09	0.09	0.08	−0.10	−0.15	0.18	1.00		
12	0.10	0.08	−0.12	0.21	0.00	0.18	0.13	0.08	−0.02	−0.07	0.04	1.00	
13	−0.05	−0.17	−0.09	0.04	−0.08	−0.04	0.06	0.15	0.07	−0.12	−0.17	−0.04	1.00

Table 8.3 Results of the Principal Component Ordination applied to the environmental data correlation matrix of the Coed Nant Lolwyn data. Only the first five rows of the transformation matrix are shown. Bold numbers show the important loadings.

Eigenvalues:

2.94	1.95	1.56	1.00	0.99	0.91	0.82	0.72	0.67	0.57	0.42	0.25	0.22

Percentage of total:

22.6	15.0	12.0	7.67	7.59	6.97	6.34	5.57	5.16	4.36	3.20	1.93	1.65

Cumulative percentage:

22.6	37.6	49.6	57.2	64.8	71.8	78.1	83.7	88.9	93.2	96.4	98.3	100.

Transformation matrix (rows are eigenvectors):

0.46	-0.01	0.11	**0.47**	-0.22	**0.51**	**0.47**	-0.06	0.06	0.10	0.03	0.15	-0.01
0.29	**-0.43**	**-0.50**	-0.28	**-0.43**	-0.01	-0.07	-0.33	-0.22	0.16	-0.02	-0.18	0.05
0.00	-0.21	-0.14	0.06	-0.16	0.01	-0.00	**0.40**	**0.40**	**-0.41**	**-0.51**	-0.03	**0.48**
-0.04	-0.30	0.21	0.11	0.10	-0.14	0.08	0.29	**-0.59**	0.22	0.07	0.19	**0.55**
0.09	-0.06	-0.01	-0.06	-0.09	-0.07	-0.14	-0.16	-0.21	**-0.46**	-0.09	**0.78**	-0.22

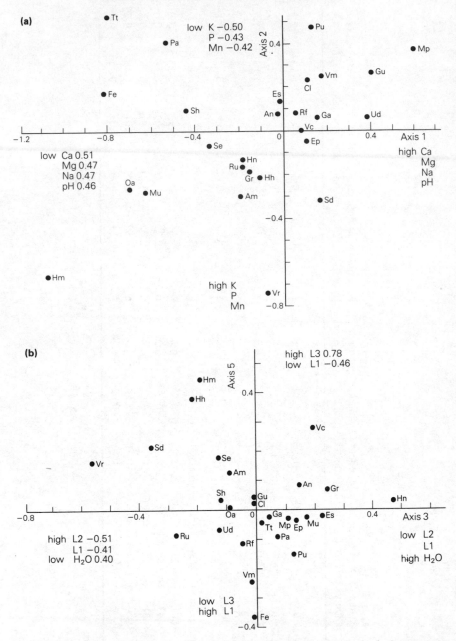

Figure 8.4 Centroid Ordination of species based on a Principal Component Ordination of the correlation matrix of the environmental factors in the Coed Nant Lolwyn case study; a Semi-direct Gradient Analysis: (a) Axes 1 and 2; (b) Axes 3 and 5. Only those species with a relative frequency of at least 10% have been shown.

particular axis, is the centroid of that species' scores in the individual stands. Thus if a species occurred with equal abundance (or were simply present in presence/absence data) in two stands only, the centroid would lie midway between the scores of those two stands on an axis. If the species' abundance values were different in the two stands, then the centroid would lie closer to the stand having the higher abundance of that species. The principle can be extended to many stands: we are essentially calculating a weighted mean of all the stand scores on each axis, the weights being the abundance values of that species in each stand (0 for absence; 1 for presence, or an abundance value for quantitative data).

The species positions in Figure 8.4 thus enable many hypotheses to be generated. *Mercurialis perennis*, *Geum urbanum* and *Urtica dioica* seem to be associated with base-rich soils, while *Holcus mollis*, *Thuidium tamariscinum* and *Oxalis acetosella* are associated with base-poor soils (Fig. 8.4a). *Hyacinthoides non-scripta* appears to occur under low spring light levels, whereas *Viola riviniana* and *Silene dioica* are correlated with high spring light; *Holcus mollis* and *Hedera helix* occur in areas of high summer light, but *Veronica montana* is associated with low summer light (Fig. 8.4b).

Remember, however, that these statements can be nothing more than preliminary hypotheses. More detailed work may continue to support some of these ideas, but not in every case.

Environmental factors in associations

If one or more environmental factors have been measured in each of a number of sample stands, and the stands have been grouped into associations by a classification method, then it is of interest to characterize the environment of each association.

An obvious way of doing this is to take one environmental factor at a time, and calculate its mean value in each of the associations. The results could be either tabulated, or plotted as a graph (see Fig. 9.2). In order to gain some idea about the reality of the differences between the mean levels of the environmental factor among the associations, a single classification analysis of variance (ANOVA) may be used, followed by the calculation of a least significant range which may be drawn on the graph. Example 8.9 shows how this is done.

Example 8.9

The distribution of soil moisture levels between the associations produced by the Information Statistic classification of the Iping Common data will be examined. The classification hierarchical diagram is shown in Figure 9.1,

Table 8.4 Tables for the Duncan multiple range test (after Duncan 1955, by permission of *Biometrics*).

5% probability level

Residual degrees of freedom	Number of groups amongst which comparisons are being made															
	2	3	4	5	6	7	8	9	10	12	14	16	18	20	50	100
1	18.0	18.0	18.0	18.0	18.0	18.0	18.0	18.0	18.0	18.0	18.0	18.0	18.0	18.0	18.0	18.0
2	6.09	6.09	6.09	6.09	6.09	6.09	6.09	6.09	6.09	6.09	6.09	6.09	6.09	6.09	6.09	6.09
3	4.50	4.50	4.50	4.50	4.50	4.50	4.50	4.50	4.50	4.50	4.50	4.50	4.50	4.50	4.50	4.50
4	3.93	4.01	4.02	4.02	4.02	4.02	4.02	4.02	4.02	4.02	4.02	4.02	4.02	4.02	4.02	4.02
5	3.64	3.74	3.79	3.83	3.83	3.83	3.83	3.83	3.83	3.83	3.83	3.83	3.83	3.83	3.83	3.83
6	3.46	3.58	3.64	3.68	3.68	3.68	3.68	3.68	3.68	3.68	3.68	3.68	3.68	3.68	3.68	3.68
7	3.35	3.47	3.54	3.58	3.60	3.61	3.61	3.61	3.61	3.61	3.61	3.61	3.61	3.61	3.61	3.61
8	3.26	3.39	3.47	3.52	3.55	3.56	3.56	3.56	3.56	3.56	3.56	3.56	3.56	3.56	3.56	3.56
9	3.20	3.34	3.41	3.47	3.50	3.52	3.52	3.52	3.52	3.52	3.52	3.52	3.52	3.52	3.52	3.52
10	3.15	3.30	3.37	3.43	3.46	3.47	3.47	3.47	3.47	3.47	3.47	3.47	3.47	3.48	3.48	3.48
11	3.11	3.27	3.35	3.39	3.43	3.44	3.45	3.46	3.46	3.46	3.46	3.46	3.47	3.48	3.48	3.48
12	3.08	3.23	3.33	3.36	3.40	3.42	3.44	3.44	3.46	3.46	3.46	3.46	3.47	3.48	3.48	3.48
13	3.06	3.21	3.30	3.35	3.38	3.41	3.42	3.44	3.45	3.45	3.46	3.46	3.47	3.47	3.47	3.47
14	3.03	3.18	3.27	3.33	3.37	3.39	3.41	3.42	3.44	3.45	3.46	3.46	3.47	3.47	3.47	3.47
15	3.01	3.16	3.25	3.31	3.36	3.38	3.40	3.42	3.43	3.44	3.45	3.46	3.47	3.47	3.47	3.47
16	3.00	3.15	3.23	3.30	3.34	3.37	3.39	3.41	3.43	3.44	3.45	3.46	3.47	3.47	3.47	3.47
17	2.98	3.13	3.22	3.28	3.33	3.36	3.38	3.40	3.42	3.44	3.45	3.46	3.47	3.47	3.47	3.47
18	2.97	3.12	3.21	3.27	3.32	3.35	3.37	3.39	3.41	3.43	3.45	3.46	3.47	3.47	3.47	3.47
19	2.96	3.11	3.19	3.26	3.31	3.35	3.37	3.39	3.41	3.43	3.44	3.46	3.47	3.47	3.47	3.47
20	2.95	3.10	3.18	3.25	3.30	3.34	3.36	3.38	3.40	3.43	3.44	3.46	3.46	3.47	3.47	3.47
22	2.93	3.08	3.17	3.24	3.29	3.32	3.35	3.37	3.39	3.42	3.44	3.45	3.46	3.47	3.47	3.47
24	2.92	3.07	3.15	3.22	3.28	3.31	3.34	3.37	3.38	3.41	3.44	3.45	3.46	3.47	3.47	3.47
26	2.91	3.06	3.14	3.21	3.27	3.30	3.34	3.36	3.38	3.41	3.43	3.45	3.46	3.47	3.47	3.47
28	2.90	3.04	3.13	3.20	3.26	3.30	3.33	3.35	3.37	3.40	3.43	3.45	3.46	3.47	3.47	3.47
30	2.89	3.04	3.12	3.20	3.25	3.29	3.32	3.35	3.37	3.40	3.43	3.44	3.46	3.47	3.47	3.47
40	2.86	3.01	3.10	3.17	3.22	3.27	3.30	3.33	3.35	3.39	3.42	3.44	3.46	3.47	3.47	3.47
60	2.83	2.98	3.08	3.14	3.20	3.24	3.28	3.31	3.33	3.37	3.40	3.43	3.45	3.47	3.48	3.48
100	2.80	2.95	3.05	3.12	3.18	3.22	3.26	3.29	3.32	3.36	3.40	3.42	3.45	3.47	3.53	3.53
∞	2.77	2.92	3.02	3.09	3.15	3.19	3.23	3.26	3.29	3.34	3.38	3.41	3.44	3.47	3.61	3.67

1% probability level

Residual degrees of freedom	Number of groups amongst which comparisons are being made															
	2	3	4	5	6	7	8	9	10	12	14	16	18	20	50	100
1	90.0	90.0	90.0	90.0	90.0	90.0	90.0	90.0	90.0	90.0	90.0	90.0	90.0	90.0	90.0	90.0
2	14.0	14.0	14.0	14.0	14.0	14.0	14.0	14.0	14.0	14.0	14.0	14.0	14.0	14.0	14.0	14.0
3	8.26	8.5	8.6	8.7	8.8	8.9	8.9	9.0	9.0	9.0	9.1	9.2	9.3	9.3	9.3	9.3
4	6.51	6.8	6.9	7.0	7.1	7.1	7.2	7.2	7.3	7.3	7.4	7.4	7.5	7.5	7.5	7.5
5	5.70	5.96	6.11	6.18	6.26	6.33	6.40	6.44	6.5	6.6	6.6	6.7	6.7	6.8	6.8	6.8
6	5.24	5.51	5.65	5.73	5.81	5.88	5.95	6.00	6.0	6.1	6.2	6.2	6.3	6.3	6.3	6.3
7	4.95	5.22	5.37	5.45	5.53	5.61	5.69	5.73	5.8	5.8	5.9	5.9	6.0	6.0	6.0	6.0
8	4.74	5.00	5.14	5.23	5.32	5.40	5.47	5.51	5.5	5.6	5.7	5.7	5.8	5.8	5.8	5.8
9	4.60	4.86	4.99	5.08	5.17	5.25	5.32	5.36	5.4	5.5	5.5	5.6	5.7	5.7	5.7	5.7
10	4.48	4.73	4.88	4.96	5.06	5.13	5.20	5.24	5.28	5.36	5.42	5.48	5.54	5.55	5.55	5.55
11	4.39	4.63	4.77	4.86	4.94	5.01	5.06	5.12	5.15	5.24	5.28	5.34	5.38	5.39	5.39	5.39
12	4.32	4.55	4.68	4.76	4.84	4.92	4.96	5.02	5.07	5.13	5.17	5.22	5.24	5.26	5.26	5.26
13	4.26	4.48	4.62	4.69	4.74	4.84	4.88	4.94	4.98	5.04	5.08	5.13	5.14	5.15	5.15	5.15
14	4.21	4.42	4.55	4.63	4.70	4.78	4.83	4.87	4.91	4.96	5.00	5.04	5.06	5.07	5.07	5.07
15	4.17	4.37	4.50	4.58	4.64	4.72	4.77	4.81	4.84	4.90	4.94	4.97	4.99	5.00	5.00	5.00
16	4.13	4.34	4.45	4.54	4.60	4.67	4.72	4.76	4.79	4.84	4.88	4.91	4.93	4.94	4.94	4.94
17	4.10	4.30	4.41	4.50	4.56	4.63	4.68	4.72	4.75	4.80	4.83	4.86	4.88	4.89	4.89	4.89
18	4.07	4.27	4.38	4.46	4.53	4.59	4.64	4.68	4.71	4.76	4.79	4.82	4.84	4.85	4.85	4.85
19	4.05	4.24	4.35	4.43	4.50	4.56	4.61	4.64	4.67	4.72	4.76	4.79	4.81	4.82	4.82	4.82
20	4.02	4.22	4.33	4.40	4.47	4.53	4.58	4.61	4.65	4.69	4.73	4.76	4.78	4.79	4.79	4.79
22	3.99	4.17	4.28	4.36	4.42	4.48	4.53	4.57	4.60	4.65	4.68	4.71	4.74	4.75	4.75	4.75
24	3.96	4.14	4.24	4.33	4.39	4.44	4.49	4.53	4.57	4.62	4.64	4.67	4.70	4.72	4.74	4.74
26	3.93	4.11	4.21	4.30	4.36	4.41	4.46	4.50	4.53	4.58	4.62	4.65	4.67	4.69	4.73	4.73
28	3.91	4.08	4.18	4.28	4.34	4.39	4.43	4.47	4.51	4.56	4.60	4.62	4.65	4.67	4.72	4.72
30	3.89	4.06	4.16	4.22	4.32	4.36	4.41	4.45	4.48	4.54	4.58	4.61	4.63	4.65	4.71	4.71
40	3.82	3.99	4.10	4.17	4.24	4.30	4.34	4.37	4.41	4.46	4.51	4.54	4.57	4.59	4.69	4.69
60	3.76	3.92	4.03	4.12	4.17	4.23	4.27	4.31	4.34	4.39	4.44	4.47	4.50	4.58	4.66	4.66
100	3.71	3.86	3.98	4.06	4.11	4.17	4.21	4.25	4.29	4.35	4.38	4.42	4.45	4.48	4.64	4.65
∞	3.64	3.80	3.90	3.98	4.04	4.09	4.14	4.17	4.20	4.26	4.31	4.34	4.38	4.41	4.60	4.68

and it will be noted that the number of stands per association ranges from 2 to 20. The mean level of soil moisture in each association is given in the calculations below, and are plotted in Figure 9.2b. The ANOVA computations are also shown below,

Sum of all the 70 values = 2395

Sum of squares of all the 70 values = 125 251

Correction factor = $(2395)^2/70$ = 81 943.21428

Association totals: A 107, B 393, C 24, D 297, E 253, F 953, G 368.

Total sum of squares = 43 307.7858

$$\text{Association sum of squares} = \frac{(107)^2}{5} + \frac{(393)^2}{13} + \frac{(24)^2}{2} + \frac{(297)^2}{20} + \frac{(253)^2}{8}$$

$$+ \frac{(953)^2}{13} + \frac{(368)^2}{9} - \text{correction factor} = 29\,836.1949$$

The ANOVA table thus appears:

	Sum of squares	Degrees of freedom	Mean square	F
Association	29 836.1949	6	4972.6992	23.26***
Residual	13 471.5909	63	213.8348	—
Total	43 307.7858	69	—	—

The association F-ratio is very highly significant, which shows that the null hypothesis of no difference between the mean levels of soil moisture in the different associations can be rejected. To compute the least significant range, we proceed as follows.

Standard error of an association mean soil moisture level
$$= \sqrt{(213.8348/10)} = 4.6242$$

With 7 associations in 70 stands, the mean number of stands per association is 10: hence the value of the denominator in the above calculation. The standard error has now to be multiplied by a factor, taken from a special table, Table 8.4 (Duncan 1955); in this instance, the appropriate value at $P(0.05)$ is 3.24. The product of 4.6242 and 3.24 is 14.98, which is the least significant range, and is plotted as the vertical bar in Figure 9.2b.

Because the number of stands per association differs so greatly, the least significant range can only be a very crude indicator of which associations differ significantly in their soil moisture contents.

The ANOVA technique can only deal with one environmental factor at a time. To consider several environmental factors together requires the multivariate extension of the analysis of variance (MANOVA) or, better still, the multivariate technique of Canonical Variate Analysis in which the associations are ordinated relative to graphical axes which themselves can be interpreted in a similar way to those of PCA. Canonical Variate Analysis thus does for whole associations what Detrended Correspondence Analysis does for individual stands; but the method involved is beyond the scope of this book.

9

Case studies – analyses

The case studies have been comprehensively introduced in Chapter 4. There, the habitats were described, and an account was given of the field methods used and of the preliminary results obtained. In this chapter we will employ some of the methods of vegetation and environmental analysis described in Chapters 6–8.

Iping Common

Divisive monothetic classification: Macnaughton-Smith Information Statistic

NORMAL

The hierarchical diagram is shown in Figure 9.1, and Figures 9.2 and 9.3 show the means of each environmental factor in each association for those factors showing significant differences between associations (see ANOVA F-ratios in Table 9.10, column 4). Table 9.1 lists the stands in each association. *Calluna vulgaris* is the first divisor species, separating out both the wet heath stands and the dense bracken zone stands, which do not contain *Calluna vulgaris*, from the remaining stands which do contain this species. Among the *Calluna*-containing stands, the first division is on *Molinia caerulea* and separates out those stands principally along the first 34 m and last 4 m which do not contain *Molinia caerulea* and in which soil moisture, organic matter and nutrient levels are low – Associations C and D (Figs 9.2b,e; Fig. 9.3).

Whether the split on *Betula pendula* into Associations C and D is ecologically meaningful is debatable. We have to bear in mind that the vegetation under scrutiny is no more than 8 years old, and that birch seeds are very small, plentiful, and easily distributed. In these circumstances one might expect young birches to occur in many different places regardless of the variations in micro-environments. If this is the case, then the difference between Associations C and D is artificial. On the other hand, it was visually apparent that even birches which were little more than seedlings were not randomly scattered, but were well zoned. Evidently there is some

244

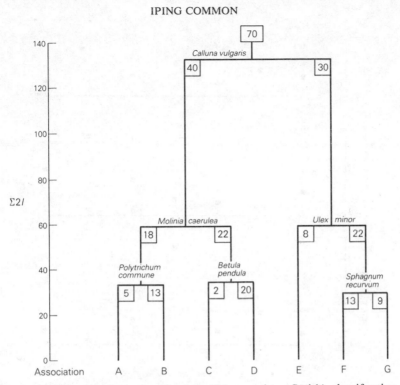

Figure 9.1 Normal Information Statistic (Macnaughton-Smith) classification of the Iping Common transect.

factor controlling the occurrence of birch other than those environmental factors included in our survey. Also, if one believes that the Information Statistic classification scheme is really ecologically sound, then the split on *Betula pendula* at a relatively high level of $\Sigma 2I$ should be regarded as defining two real species assemblages.

Similar remarks may be made in relation to the split on *Polytrichum commune*, giving Associations A and B, but here the stands of the former have markedly lower soil calcium, magnesium and potassium levels than the latter (Figs 9.3a,b,c; 4.2a,b,d). Although this moss is a conspicuous feature of Stands 114–120 m, where it occurs with high cover values, only Stands 114 and 115 occur in Association A. In other words, only Stand 114 and 115 contain *Polytrichum commune* and *Molinia caerulea* and *Calluna vulgaris*. Stands 116, 118 and 120 do not contain *Molinia caerulea*, but they do contain both *Polytrichum commune* and *Calluna vulgaris*. *Polytrichum commune* also occurs in Association G in Stands 79, 80 and 82, where *Calluna vulgaris* is not found. It is thus important to remember that a divisor species (apart from the primary one) can occur in other associations which are not defined by the divisor species concerned.

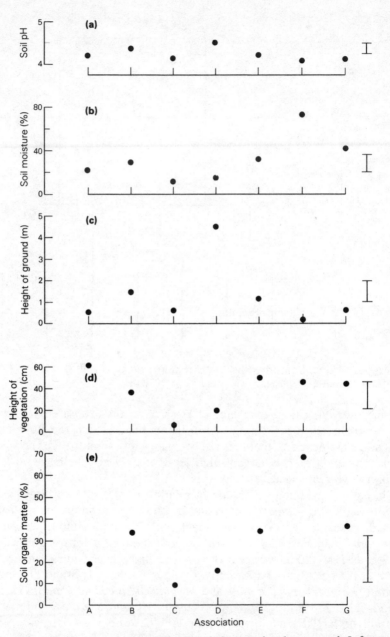

Figure 9.2 Mean values of environmental factors in the normal Information Statistic associations of the Iping Common transect: (a) soil pH; (b) soil moisture; (c) height of ground above lowest point on the transect; (d) height of vegetation; (e) soil organic matter. The approximate least significant ranges are at $P(0.05)$.

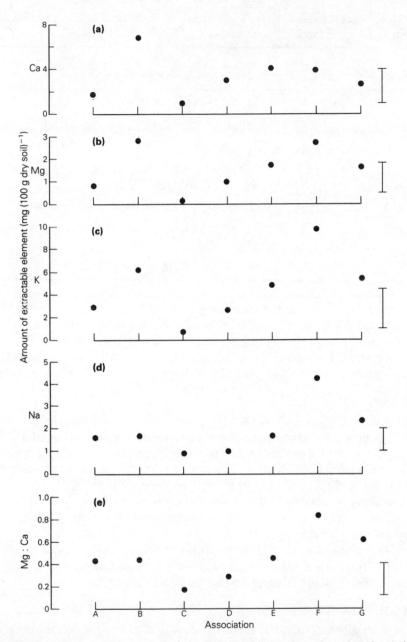

Figure 9.3 Mean values of environmental factors in the normal Information Statistic associations of the Iping Common transect: extractable soil (a) calcium; (b) magnesium; (c) potassium; (d) sodium; (e) magnesium : calcium ratio. The approximate least significant ranges are at $P(0.05)$.

Table 9.1 List of stands in each association of the normal Information Statistic Analysis of the Iping Common transect.

Association	Number of stands	List of stands
A	5	68, 75, 77, 114, 115
B	13	36, 38, 42, 44, 46, 48, 52, 54, 70, 72, 73, 76, 78
C	2	118, 120
D	20	0, 2, 4, 6, 8, 10, 12, 14, 16, 18, 20, 22, 24, 26, 28, 30, 32, 34, 40, 116
E	8	50, 55, 56, 57, 58, 66, 69, 74
F	13	84, 86, 88, 90, 92, 94, 96, 98, 100, 104, 106, 108, 110
G	9	60, 62, 64, 79, 80, 82, 102, 112, 113

Among the 30 stands not containing *Calluna vulgaris*, eight contain *Ulex minor* and form Association E. Out of the remaining stands, 13 include *Sphagnum recurvum* – the boggy part of the transect (Association F), while Association G consists of miscellaneous stands of relatively high moisture and organic contents, and generally high soil nutrients (Figs 9.2b,e; 9.3).

INVERSE

The inverse analysis is shown in Figure 9.4, and Table 9.2 lists the species in each association. Among the 5 associations of species extracted by the analysis, the first 4 are significant in terms of ecological affinities, and the last association is essentially a collection of miscellaneous species – the 'rag bag' (see p. 281, but here read 'species' for 'stand' and vice versa). Association A comprises those species occurring on damp ground with a moderate level of nutrients, while Association B comprises the species of dry ground and low nutrient levels. Association C is the definitive set of species of the wet heath, while the species of Association D occur principally in Associations A and C of the normal analysis, the stands of which tend to occur at equal heights on either side of the wet heath basin.

NODAL ANALYSIS

The Nodal Analysis, shown in Table 9.3, displays the whole data set analyzed both ways, and certainly helps to put things into perspective. Disregarding the miscellaneous species of Inverse Association E, the remaining normal and inverse associations have been arranged to give a diagonal of species-in-stand occurrences from the top left to the bottom

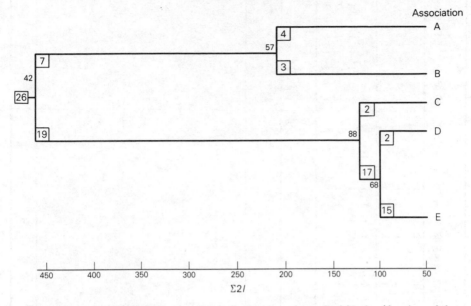

Figure 9.4 Inverse Information Statistic (Macnaughton-Smith) classification of the Iping Common transect.

Table 9.2 List of species in each association of the inverse Information Statistic Analysis of the Iping Common transect.

Association	Number of species	List of species
A	4	*Erica tetralix, Ulex minor, Molinia caerulea, Pteridium aquilinum*
B	3	*Calluna vulgaris, Erica cinerea, Campylopus interoflexus*
C	2	*Juncus effusus, Sphagnum recurvum*
D	2	*Betula pendula, Polytrichum commune*
E	15	*Betula pubescens, Quercus robur, Ulex europaeus, Polytrichum juniperinum, Sphagnum tenellum, Cladonia coccifera, Cladonia crispata, Cladonia coniocraea, Cladonia fimbriata, Cladonia floerkiana, Hypogymnia physodes, Aulacomnium palustre, Webera nutans, Carex nigra, Juncus bulbosus*

249

Table 9.3 Nodal Analysis, based on Information Statistic classifications, of the Iping Common transect.

	0	2	4	6	8	10	12	14	16	18	20	22	24	26	28	30	32	34	40	36	38	42	44	46	48	52	54	70	72	73	76	78	50	55	56	57	58	66	69	74
Calluna vulgaris	1	1	1	1	1	1	1	1	1	1	1	1	1	1	1	1	1	1	1	1	1	1	1	1	1	1	1	1	1	1	1	1								1
Erica cinerea	1	1	1	1	1	1	1	1	1	1	1	1	1	1	1	1	1	1	1	1	1	1	1	1	1	1	1	1	1	1	1	1								
Campylopus interofl.	1	1	1	1	1	1	1	1	1	1	1	1	1	1	1	1	1	1	1	1	1	1	1	1	1	1	1	1	1	1	1	1								
Erica tetralix															1			1		1	1	1	1	1	1	1	1	1	1	1	1	1	1	1	1	1	1	1	1	1
Ulex minor																		1	1	1	1	1	1	1	1	1	1	1	1	1	1	1	1	1	1	1	1	1	1	1
Molinia caerulea													1	1					1	1	1	1	1	1	1	1	1	1	1	1	1	1	1	1	1	1	1	1	1	1
Pteridium aquilinum																			1	1	1	1	1	1	1	1	1	1	1	1	1	1	1	1	1	1	1	1	1	1
Betula pendula																																								1
Polytrichum commune																																								
Juncus effusus																																								
Sphagnum recurvum																																								
Betula pubescens																																1								
Quercus robur																																								
Ulex europaeus																								1	1															
Polytrichum juniper.					1																																			
Sphagnum tenellum																																								
Cladonia coccifera			1	1	1																																			
Cladonia crispata			1	1																																				
Cladonia coniocraea				1																																				
Cladonia fimbriata				1																																				
Cladonia floerkiana	1	1																																						
Hypogymnia physodes	1	1	1	1												1																								
Aulacomnium palustre															1																									
Webera nutans													1																											
Carex nigra																																								
Juncus bulbosus																																								
Normal association	D																			B													E							

Table (rows = species; columns = stand/sample numbers grouped by "Normal association" A, C, G, F; right-hand column = "Inverse association" B, A, D, C, E):

Species	68	75	77	114	115	118	120	60	62	64	79	80	82	102	112	113	84	86	88	90	92	94	96	98	100	104	106	108	110	Inverse association
Calluna vulgaris	1	1	1	1	1	1	1																							B
Erica cinerea					1	1	1																							B
Campylopus interofl.																														B
Erica tetralix	1			1		1					1					1												1	1	A
Ulex minor	1	1	1	1	1			1	1	1	1	1	1	1	1	1												1	1	A
Molinia caerulea				1	1			1	1	1	1	1	1	1	1	1									1	1	1	1	1	A
Pteridium aquilinum	1							1	1	1							1	1	1	1	1	1	1	1	1	1	1	1	1	A
Betula pendula	1	1	1	1	1	1	1					1			1	1														D
Polytrichum commune	1	1	1	1	1	1	1				1	1			1	1														D
Juncus effusus				1							1												1		1	1				C
Sphagnum recurvum																	1	1	1	1	1	1	1	1	1	1	1	1	1	C
Betula pubescens					1	1	1																							E
Quercus robur						1	1																							E
Ulex europaeus			1							1																				E
Polytrichum juniper.														1									1							E
Sphagnum tenellum																														E
Cladonia coccifera																														E
Cladonia crispata																														E
Cladonia coniocraea																														E
Cladonia fimbriata																														E
Cladonia floerkiana																														E
Hypogymnia physodes																														E
Aulacomnium palustre																					1									E
Webera nutans													1	1																E
Carex nigra												1																		E
Juncus bulbosus																														E
Normal association	A					C		G									F													

right of the table. In the top left-hand corner we find the stands of Association D mostly containing the species of Association B. These are the *Calluna*-dominated stands at the top of the transect; but as we go down the line, species of Association A start appearing in the stands. In the group of columns second from the left (stand Association B) the species of Association A are much more in evidence, and those of Association B start to decline in frequency. In the third group of columns (stand Association E), the species of Association B have almost disappeared and those of Association A predominate. These first three groups of columns represent a progression down the line from the highest point.

The groups of columns at the right-hand end of the table contains the wet heath stands (Association F), defined by *Sphagnum recurvum*, and the three groups of columns to the left (stand Associations A, C and G) contain stands on either side of the wet heath basin.

Most of the species of Association E, the miscellaneous set, are very scattered in the stand associations. The exception to this is the group of lichen species which demonstrate a greater coherence.

Divisive polythetic classification: Indicator Species Analysis (qualitative)

NORMAL

Since the normal Information Statistic has produced 7 associations, we specify 3 levels of division here which produce 8 associations. The hierarchical diagram is shown in Figure 9.5, and mean values of environmental factors in each association in Figures 9.6 and 9.7. Table 9.4 lists the stands in each association. In the hierarchical diagram, the level of each division has been placed at the appropriate point on a vertical scale of eigenvalues of the first ordination axis.

The first split is into 15 stands of essentially the wet heath from 80 to 104 m with *Sphagnum recurvum* and *Juncus effusus* as indicator species. Stands 82 and 102 m are excluded, however, since they do not contain the indicator species (see Nodal Analysis, Table 9.6). The next split on this positive side is on *Juncus effusus* as indicator species; in fact, Table 9.6 shows that the split is made entirely on the presence (+ ve side) and absence (– ve side) of this species. Since *Juncus effusus* is a tall plant, and dominates in most of the stands in which it occurs, this division seems ecologically, as well as visually, sensible.

Of the 7 stands not containing *Juncus effusus* there is a division on *Erica tetralix*: 2 stands contain this species, and 5 stands do not. Is this an ecologically relevant division? Association E, containing *Erica tetralix*, does have conspicuously higher mean soil organic matter, potassium, and

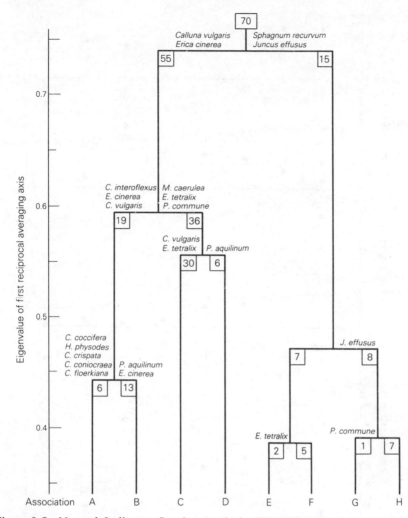

Figure 9.5 Normal Indicator Species Analysis (TWINSPAN) of the qualitative data of the Iping Common transect.

manganese contents, also a higher magnesium:calcium ratio (Figs 9.6d; 9.7a,c,d) than Association F (not containing *Erica tetralix*); but of these 4 factors only soil organic matter and manganese are consistently lower in the stands of Association F than in those of Association E (see Figs 4.1b; 4.2c, d; 4.3b). What about the one stand of Association G, split from the 7 stands of Association H? Stand 80 m is the only one of the 15 on the positive side of the initial dichotomy to contain *Polytrichum commune*, and also has low values of soil moisture, organic matter, and inorganic ions

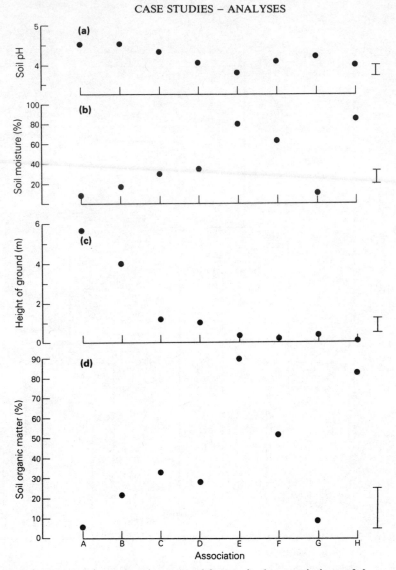

Figure 9.6 Mean values of environmental factors in the associations of the normal Indicator Species Analysis of the qualitative data of the Iping Common transect: (a) soil pH; (b) soil moisture; (c) height of ground; (d) soil organic matter.

generally (Figs 4.1b; 4.2; 4.3; 9.6b,d; 9.7a,b,c). Stand 80 m is obviously quite different from those of Association H; it has close affinities with Stand 115 m, but the latter contains *Calluna vulgaris* and *Erica cinerea* also. It is the absence of these species which cause Stand 80 m to be included on the right-hand side of the main dichotomy.

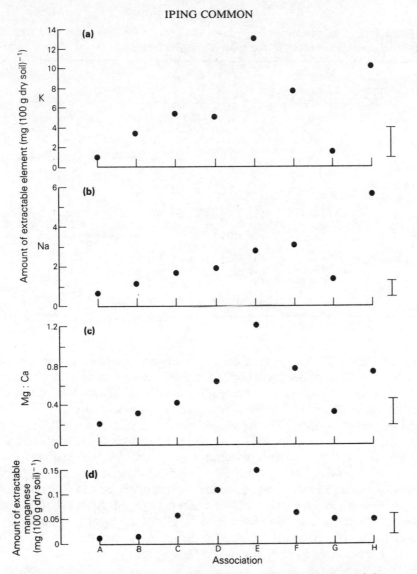

Figure 9.7 Mean values of environmental factors in the associations of the normal Indicator Species Analysis of the qualitative data of the Iping Common transect: extractable soil (a) potassium; (b) sodium; (c) magnesium : calcium; (d) manganese.

Turning now to the 55 stands on the negative side, the first division is into 19 stands which mostly contain *Calluna vulgaris*, *Erica cinerea*, and *Campylopus interoflexus* – dry and of low nutrient status – and 36 stands of damper, somewhat richer soil in which *Molinia caerulea*, *Erica tetralix* and, to a lesser extent, *Polytrichum commune* occur. Of the first 19 stands

Table 9.4 List of stands in each association of the normal Indicator Species Analysis (qualitative data) of the Iping Common transect.

Association	Number of stands	List of stands
A	6	0, 2, 4, 8, 12, 14
B	13	6, 10, 16, 18, 20, 22, 24, 26, 28, 30, 32, 34, 40
C	30	36, 38, 42, 44, 46, 48, 50, 52, 54, 55, 56, 57, 68, 70, 72, 73, 74, 75, 76, 77, 78, 79, 82, 102, 113, 114, 115, 116, 118, 120
D	6	58, 60, 62, 64, 66, 69
E	2	106, 108
F	5	84, 86, 92, 110, 112
G	1	80
H	7	88, 90, 94, 96, 98, 100, 104

there is a final dichotomy into 6 stands mainly characterized by the presence of lichens (Association A), and 13 stands containing very few lichens (Association B). Other than this, only soil organic matter and potassium levels seem markedly different between the two associations.

The group of 36 stands are divided into 6, all containing *Pteridium aquilinum*, which constitute the vigorous bracken zone (Association D) and comprise little other than *Pteridium aquilinum*, *Molinia caerulea*, and often *Ulex minor*. The remaining 30 stands all constitute Association C, and this seems to be the most heterogeneous group. Even allowing one more level of division only splits off one stand (102 m). Although Indicator Species Analysis does not contain a 'rag-bag' of assorted stands as a monothetic method may do (p. 281), nevertheless Association D in the present instance comes rather close to this.

INVERSE

The hierarchical diagram is shown in Figure 9.8, and Table 9.5 gives a list of species in each association. The first dichotomy splits off the 4 species occurring on the wettest ground, and there is no further dichotomy here (Association D). Of the remaining 22 species, *Erica tetralix*, *Molinia caerulea*, and *Polytrichum commune* form Association C of damp ground species, although the last attains its maximum cover values in dry soils. Association A comprises dry ground species: *Calluna vulgaris*, *Erica*

256

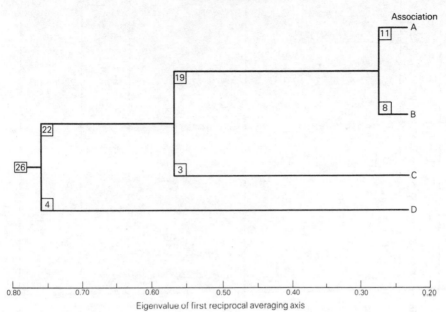

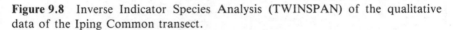

Figure 9.8 Inverse Indicator Species Analysis (TWINSPAN) of the qualitative data of the Iping Common transect.

Table 9.5 List of species in each association of the inverse Indicator Species Analysis (qualitative data) of the Iping Common transect.

Association	Number of species	List of species
A	11	*Calluna vulgaris, Erica cinerea, Polytrichum juniperinum, Cladonia coccifera, Cladonia crispata, Cladonia coniocraea, Cladonia fimbriata, Cladonia floerkiana, Hypogymnia physodes, Campylopus interoflexus, Webera nutans*
B	8	*Betula pendula, Betula pubescens, Quercus robur, Ulex europaeus, Ulex minor, Pteridium aquilinum, Aulacomnium palustre, Carex nigra*
C	3	*Erica tetralix, Molinia caerulea, Polytrichum commune*
D	4	*Juncus effusus, Sphagnum tenellum, Sphagnum recurvum, Juncus bulbosus*

Table 9.6 Nodal Analysis, based on qualitative Indicator Species Analysis classifications, of the Iping Common transect.

	0	2	4	8	12	14	6	10	16	18	20	22	24	26	28	30	32	34	40	36	38	42	44	46	48	50	52	54	55	56	57	68	70	72	73	74	75	76	77	78
Calluna vulgaris	1	1	1	1	1	1	1	1	1	1	1	1	1	1	1	1	1	1	1	1	1	1	1	1	1	1	1	1	1	1	1	1	1	1	1	1	1	1	1	1
Erica cinerea	1	1				1	1	1	1	1	1	1	1	1	1	1	1	1	1	1	1	1	1	1	1	1	1	1	1	1		1								
Polytrichum juniper.			1	1	1	1		1																																
Cladonia coccifera			1	1	1	1			1																															
Cladonia crispata			1		1																																			
Cladonia coniocraea			1	1		1																																		
Cladonia fimbriata				1	1	1																																		
Cladonia floerkiana			1	1																																				
Hypogymnia physodes					1	1	1	1												1				1	1							1				1				
Campylopus interofl.	1	1	1	1	1	1	1	1			1	1	1	1	1	1	1	1	1	1	1	1	1	1	1	1	1	1	1											
Webera nutans											1	1	1	1	1	1	1	1	1	1	1				1		1	1												
Betula pendula																						1									1	1				1				
Betula pubescens																																		1						
Quercus robur																											1													
Ulex europaeus																				1													1			1				
Ulex minor																			1	1	1	1	1	1	1	1	1	1	1	1	1	1	1			1				
Pteridium aquilinum													1	1			1	1	1	1	1	1	1	1	1	1	1	1	1	1	1	1	1							
Aulacomnium palustre													1						1																					
Carex nigra															1				1								1													
Erica tetralix																				1	1	1	1	1	1	1	1	1	1	1	1	1	1	1	1	1	1	1	1	1
Molinia caerulea																				1	1	1	1	1	1	1	1	1	1	1	1	1	1	1	1	1	1	1	1	1
Polytrichum commune																															1	1							1	
Juncus effusus																																								
Sphagnum tenellum																																								
Sphagnum recurvum																																								
Juncus bulbosus																																								
Normal association	A						B													C																				

Species	79	82	102	113	114	115	116	118	120	58	60	62	64	66	69	105	108	84	86	92	110	112	80	88	90	94	96	98	100	104	Inverse association
Calluna vulgaris						1	1	1	1																						
Erica cinerea						1	1	1																							
Polytrichum juniper.													1																		A
Cladonia coccifera																															
Cladonia crispata																															
Cladonia coniocraea																															
Cladonia fimbriata																															
Cladonia floerkiana																															
Hypogymnia physodes																															
Campylopus interofl.							1																								
Webera nutans																															
Betula pendula					1		1	1	1																						
Betula pubescens								1	1																						B
Quercus robur				1																											
Ulex europaeus																															
Ulex minor													1	1	1																
Pteridium aquilinum										1	1	1	1	1	1																
Aulacomnium palustre		1																													
Carex Nigra	1																														
Erica tetraix	1	1	1							1						1	1	1	1	1	1	1	1	1	1					1	
Molinia caerulea	1	1	1							1						1	1	1	1	1	1	1	1								C
Polytrichum commune	1	1	1	1				1		1								1	1	1	1	1									
Juncus effusus						1										1	1						1	1	1						
Sphagnum tenellum																1	1							1	1	1	1	1	1	1	
Sphagnum recurvum																					1	1		1	1	1	1	1	1	1	D
Juncus bulbosus																															
Normal association	C									D						E		F					G	H							

cinerea, and lichens; while Association B consists of a miscellaneous collection of damp ground species.

NODAL ANALYSIS

The Nodal Analysis is shown in Table 9.6 and requires little comment. It is now apparent, though, that stand Association C has the greatest 'concentration' of Association C species and, to a lesser extent, of Association B species.

Divisive polythetic classification: Indicator Species Analysis (quantitative)

NORMAL

Again we specify 3 levels of division, and the hierarchical diagram is shown in Figure 9.9, and a list of stands in each association is given in Table 9.7. The first division is almost identical to that of the qualitative Indicator Species Analysis: the same indicator species are involved but one stand, at 102 m, is now on the positive side instead of the negative. Three of the four indicator species are at abundance level 1; but *Juncus effusus* is at level 3, showing that Stand 115 m containing this species at abundance level 2 is not sufficient to place that stand on the positive side of the initial dichotomy. Indeed, a glance at the Nodal Analysis table (Table 9.9) shows that the floristic composition of Stand 115 m is quite different from the other stands in which *Juncus effusus* occurs; *Juncus effusus* is an outlier species in Stand 115 m.

The next division on the positive side is very similar to the corresponding division in the qualitative Indicator Species Analysis, but is rather more explicit about the indicator species. Thus, only abundant *Molinia caerulea* (levels 4 & 5) is indicative of the negative side of this dichotomy, but the mere presence of *Juncus effusus* suffices on the positive side. Further down on the positive side, Stand 80 m is again singled out on the basis of the presence of *Polytrichum commune* as Association F, and the remaining 8 stands not containing this species form Association G. The 7 stands on the negative side of the second level split are further subdivided differently from the qualitative case, and form Associations D and E. Association E is defined by the presence of at least a moderate cover of *Sphagnum recurvum*, but does not seem to be environmentally different from Association D.

The negative side of the initial dichotomy subdivides differently from that of the qualitative Indicator Species Analysis. From a vegetation abundance viewpoint, Stand 12 m was very conspicuous for its lack of tall vegetation, so it is not surprising that this stand is 'pulled out' high up in the sequence.

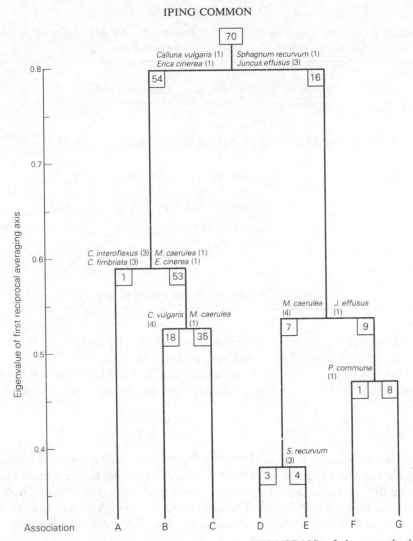

Figure 9.9 Normal Indicator Species Analysis (TWINSPAN) of the quantitative data of the Iping Common transect.

The remaining 53 stands are essentially divided into stands containing *Molinia caerulea* (Association C) and those not containing this species (Association B). Further, all stands of Association B contain at least 25% cover of *Calluna vulgaris*, but many stands of Association C contain less of this species (see Table 9.9). Both these groups are rather large, and a fourth level of division was examined. Association B split off the four stands at 24, 26, 34, 40 m, i.e. stands containing *Pteridium aquilinum* (Table 9.9), while Association C split off Stands 60, 62, 64, 66, 69 m – again containing

Table 9.7 List of stands in each association of the normal Indicator Species Analysis (quantitative data) of the Iping Common transect.

Association	Number of stands	List of stands
A	1	12
B	18	0, 2, 4, 6, 8, 10, 14, 16, 18, 20, 22, 24, 26, 28, 30, 32, 34, 40
C	35	36, 38, 42, 44, 46, 48, 50, 52, 54, 55, 56, 57, 58, 60, 62, 64, 66, 68, 69, 70, 72, 73, 74, 75, 76, 77, 78, 79, 82, 113, 114, 115, 116, 118, 120
D	3	102, 110, 112
E	4	84, 86, 106, 108
F	1	80
G	8	88, 90, 92, 94, 96, 98, 100, 104

bracken, but only when this species has high cover (Table 9.9); in other words, these 5 stands have dense and vigorous bracken.

Hardly any useful new information is provided by the distribution of the environmental factor means in the associations, so these results are not shown.

INVERSE

The hierarchical diagram is shown in Figure 9.10, and a list of species in each association is given in Table 9.8. There is little difference between the qualitative and quantitative analyses in respect of Associations A and B, but *Molinia caerulea* now forms an association of its own. Association F in the present case is very similar to Association D of the qualitative analysis.

NODAL ANALYSIS

The Nodal Analysis appears in Table 9.9, but requires no further comment.

Comparison of the classifications (normal)

THE DISTRIBUTIONS OF ASSOCIATIONS ALONG THE TRANSECT

In comparing the three normal, or stand, classifications, a map of the transect stands showing the results of the classifications will be useful: it appears in Figure 9.11. Remember that groupings of stands which are

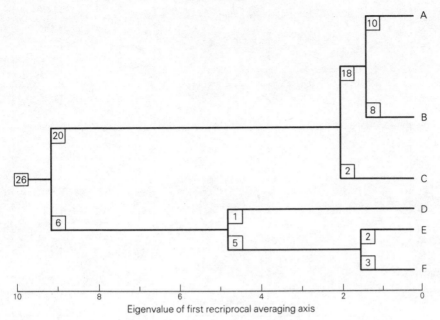

Figure 9.10 Inverse Indicator Species Analysis (TWINSPAN) of the quantitative data of the Iping Common transect.

Table 9.8 List of species in each association of the inverse Indicator Species Analysis (quantitative data) of the Iping Common transect.

Association	Number of species	List of species
A	10	*Calluna vulgaris, Erica cinerea, Cladonia coccifera, Cladonia crispata, Cladonia coniocraea, Cladonia fimbriata, Cladonia floerkiana, Hypogymnia physodes, Campylopus interoflexus, Webera nutans*
B	8	*Betula pendula, Betula pubescens, Quercus robur, Ulex europaeus, Ulex minor, Pteridium aquilinum, Polytrichum juniperinum, Carex nigra*
C	2	*Erica tetralix, Polytrichum commune*
D	1	*Molinia caerulea*
E	2	*Sphagnum tenellum, Aulacomnium palustre*
F	3	*Juncus effusus, Sphagnum recurvum, Juncus bulbosus*

263

Table 9.9 Nodal Analysis, based on quantitative Indicator Species Analysis classifications, of the Iping Common transect.

Species	12	0	2	4	6	8	10	14	16	18	20	22	24	26	28	30	32	34	40	36	38	42	44	46	48	50	52	54	55	56	57	58	60	62	64	66	68	69	70	72
Calluna vulgaris	1	5	5	5	4	4	5	4	5	5	5	5	4	5	5	5	5	5	4	3	4	3	2	4	5	3	3	2	5	3	3	3	3	3	3		2		4	3
Erica cinerea		3	3	2	1			2		2		3	3	3	4	5	3	1	3	3	4	3	2	4		2	1	1	5	3	5			5	3					3
Cladonia coccifera		2	1	1		1																																		
Cladonia crispata		2		1																																				
Cladonia coniocraea	3	1																																						
Cladonia fimbriata	3																																							
Cladonia floerkiana		1	1																																					
Hypogymnia physodes		1	2	1	1										1																									
Campylopus interofl.	4	3	1	3	1	3	3		3		3	2	1	3		4	4	3	3	3	1	1		3		2	3													
Webera nutans													1	1																										
Betula pendula																																			5					
Betula pubescens																																								
Quercus robur																							4	4																
Ulex europaeus																																								
Ulex minor												5	3					1	4	4	3	3		4	2	4		1	4	5	5	3	3	3	4		2		5	
Pteridium aquilinum							1									3	3	3	3	3	4	3		3	3	2	2	4	5	5	5	3	5	5	1					
Polytrichum juniper.																																3								
Carex nigra																																								
Erica tetralix														3	3	1				3	4	3	5		3	3	4	3	4	2	3						3			4
Polytrichum commune																																			2					
Molinia caerulea																				2	3	2	1	1	4	2	2	1	3	4	3	5	5		5	3	4	3	3	
Sphagnum tenellum																																								
Aulacomnium palustre																																								
Juncus effusus																																								
Sphagnum recurvum																																								
Juncus bulbosus																																								
Normal association	A										B															C														

Species	73	74	75	76	77	78	79	82	113	114	115	116	118	120	102	110	112	84	86	106	108	80	88	90	92	94	96	98	100	104	Inverse association
Calluna vulgaris	5	5	5	4					3	3	3		2	3																	A
Erica cinerea											2			3																	
Cladonia coccifera																															
Cladonia crispata																															
Cladonia coniocraea																															
Cladonia fimbriata																															
Cladonia floerkiana																															
Hypogymnia physodes																															
Campylopus interofl.	2											3																			
Webera nutans																															
Betula pendula	4			4				5		1	1			3																	B
Betula pubescens		3					3						1																		
Quercus robur								1		1																					
Ulex europaeus					3		3																								
Ulex minor	3																														
Pteridium aquilinum																															
Polytrichum juniper.																															
Carex nigra								1																							
Erica tetralix	3	4	4		4	3	4	2	4	3		5	1	5						4	4	3									C
Polytrichum commune	3	4				5	2	3				5	5	5																	
Molinia caerulea	3	4	2	5	3	2	4	5	2	3					5	5	5	5	5	5	4	4	3	1	1					5	D
Sphagnum tenellum															2																E
Aulacomium palustre																				2											
Juncus effusus												2						3	5	3	4	3	4	4	5	4	4	5	5	3	F
Sphagnum recurvum																	2						5	5	5	5	5	5	5	5	
Juncus bulbosus																									2						
Normal association									C							D			E			F				G					

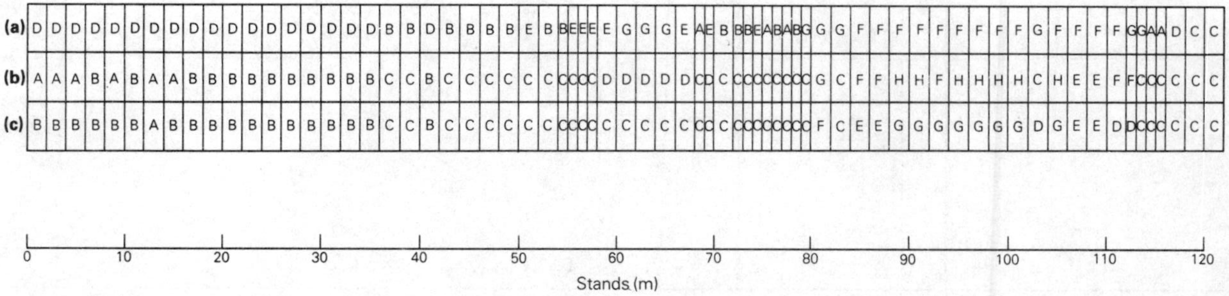

(a) D D D D D D D D D D D D D D D D B D B B B B B B BE E E G G G E AE B BBEABABG G F F F F F G F F F G GAAD C C

(b) A A A B A B A B B B B B B B C B C C C C C C D D C OC O D D D D C OC O C OC OG C F F H H C H E E F FCCC C C C

(c) B B B B B B B A B B B B B B B C B C C C C C C C OC OC OC C OC OC OC OF C E E E G G G G G G D G E E D CCCC C C

Stands (m)

0   10   20   30   40   50   60   70   80   90   100   110   120

Figure 9.11 Map of the Iping Common transect showing the distributions of the associations produced by: (a) Information Statistic classification; (b) Indicator Species Analysis, qualitative; (c) Indicator Species Analysis, quantitative.

essentially the same will usually have different letters in the different methods.

Starting from the top end of the transect, the Information Statistic and the quantitative Indicator Species Analysis show virtually the same result from 0 to 48 m, and the qualitative Indicator Species Analysis differs in only one respect. Apart from Stand 12 m, which was almost devoid of any vegetation other than lichens, Information Statistic and quantitative Indicator Species Analysis both show Stands 0 to 34 m and 40 m as one association. Qualitative Indicator Species Analysis splits off as a separate association most of the stands in the first 14 m which contain the lichen species, but otherwise this analysis agrees with the other two.

At the other end of the transect, the two Indicator Species Analyses give identical results, apart from Stands 112, 110, 102 m and 92 back to 80 m; but Information Statistic treats this part of the transect somewhat differently. Between 50 and 78 m, Indicator Species Analysis, particularly the quantitative version, indicates a considerable uniformity in the vegetation, whereas Information Statistic gives four different associations in this length.

Which of these classifications is 'best'; and what, in fact, do we mean by 'best' in this context? Let us review some of the differences in the results of the different classificatory methods and consider the importance of the various vegetational features highlighted by the different methods.

First consider Stand 12 m, the stand with very sparse vegetation (apart from lichens), which is highlighted by the quantitative Indicator Species Analysis. Of the environmental factors measured, the only differences between Stand 12 m and neighbouring ones are heightened calcium and manganese levels in that stand (Figs 4.2a, 4.3b) – conditions hardly conducive to *less* vegetation. Obviously one must look to more secondary features. Thus the stand might have been the site of a picnic fire a year or two previously, although no clear evidence of a fire was visible. So the highlighting of Stand 12 m as a separate association by quantitative Indicator Species Analysis is not very helpful. But in specifying only 3 levels of division, to give a maximum of 8 associations, having Stand 12 as one association in effect 'wastes' an association leaving only 7 to be shared out among the remaining 69 stands.

The next feature of the transect, that is separated out by one of the three methods only, is the set of lichen-containing stands at the top. Thus, qualitative Indicator Species Analysis places Stands 0, 2, 4, 8, 12, 14 m in a separate association from the remaining stands to 34 m. All these stands include lichen species, but Stands 6 and 10 m have not been included, possibly because these two stands contain only one lichen species together with *Erica cinerea* (Table 9.6). Certainly the more intermediate nature of Stands 6 and 10 m show up on the first axis of the qualitative Detrended

Correspondence Ordination (Fig. 9.18), but in classification the line of demarcation has to be put somewhere. Is the segregation of the lichen-containing stands ecologically relevant? In general, the lichen zone extended down to 14 m, and from here to 20 m the *Calluna vulgaris* appeared to be more vigorous in its growth (it was taller rather than having greatly increased cover), but there does not seem to be a concomitant increase in soil nutrients (Figs. 4.2 & 4.3). Ecologically, therefore, the distinction between the two sets of stands seems worth making, but from a purely large-scale mapping viewpoint, this dichotomy may be less relevant.

The length of transect between 42 and 79 m is undivided by the quantitative Indicator Species Analysis, but one further division splits off almost the same stands as already shown as a separate group in the qualitative Indicator Species Analysis – the tall bracken zone. Information Statistic shows four different associations over this length, but the dense bracken zone is not very clearly defined. On the whole, the fragmentation of the 42–79 m length of the transect appears to be governed by the presence and absence of *Calluna vulgaris*; so the relevance of having different associations essentially hinges on the ecological importance of *Calluna vulgaris*. There seems to be some indication of a positive correlation between *Calluna vulgaris* presence and soil calcium and magnesium *in this part of the transect* (Figs 4.2a, b & 4.4a), but nothing else emerges from our survey. More detailed work would be required to establish the ecological relevance of the presence or absence of *Calluna vulgaris* in this part of the transect.

Over the last 40 m of the transect, the two versions of Indicator Species Analysis yield the same result apart from discrepancies within a few isolated stands. Information Statistic tends to give a single association between 84 and 110 m (apart from Stand 102 m), whereas the Indicator Species Analyses have two associations, which reflect the change from dominant *Molinia caerulea* to dominant *Juncus effusus* just after Stand 86 m and the reverse at 105 m. Qualitative Indicator Species Analysis shows Stand 92 m as a different association from those on either side of it; and this stand certainly appeared to be different, being situated in a small area of little or no tall plants and 100% cover of *Sphagnum recurvum*. Again, Stands 113 to 120 m are shown as uniform in the two versions of Indicator Species Analysis, but are fragmented into four associations by the Information Statistic.

From a purely visual point of view, the qualitative Indicator Species Analysis would seem to be the best in that the important changes of species assemblages are reflected by changes of associations. On the other hand, one may ask whether these visual changes are all *ecologically* relevant; it may be that some of the associations demonstrated by Information Statistic, which are not as readily apparent to the eye, are of more ecological significance than are those changes in associations which correspond with visually obvious vegetation changes. Only more detailed work could answer this question.

THE ENVIRONMENTAL CHARACTERISTICS OF ASSOCIATIONS

If one pursues the axiom of ecology that species distribution is related to environmental factors, then it would be expected that marked differences between associations may be found in at least some of the environmental parameters measured in the study. The appropriate method of analysis was described in Chapter 8 (pp. 239–43) and consists of a single classification ANOVA applied to each environmental factor measured in the field. For the present case study, the relevant results are summarized in Table 9.10.

The number of significant F-ratios does not differ much between the three classification methods. All the methods agree on the importance of soil pH, moisture, organic matter, potassium, sodium, and the magnesium : calcium ratio in characterizing some of the associations. The remaining environmental factors are highlighted by different classification methods. Soil phosphorus is 'picked out' by quantitative Indicator Species Analysis; soil iron, calcium and magnesium, together with vegetation height, by Information Statistic; while differences in soil manganese are maximal in qualitative Indicator Species Analysis. On this basis, therefore, it would seem as though the Information Statistic is the best classification.

If the number of associations was the same for each of the three classification methods, then the order of magnitude of the residual mean squares would be the converse of the order of magnitude of the F-ratios for any one environmental factor; for example, the classification method giving the smallest residual mean square would also give the largest F-ratio.

However, our present situation is complicated by the fact that the qualitative Indicator Species Analysis has one more association (8) than the other two methods (7). Consequently, it may be better to compare the performance of the classification methods by residual mean squares, bearing in mind that a small residual mean square for a particular environmental factor denotes homogeneous associations with respect to that factor.

The Information Statistic classification gives a substantially lower residual mean square for soil calcium, and slightly lower for magnesium, than the other two classifications. The qualitative Indicator Species Analysis gives substantially lower residual mean squares for soil moisture, organic matter, potassium, sodium, manganese, and height of ground; also a slightly lower residual mean square for the magnesium : calcium ratio. The quantitative Indicator Species Analysis shows only soil phosphorus to have the lowest residual mean square. Now it would appear that the qualitative Indicator Species Analysis is the best classification from the viewpoint of

269

Table 9.10 The residual mean squares and F-ratios of the ANOVAs of the environmental factors based on associations produced by Information Statistic and Indicator Species classifications of the Iping Common data (both qualitative and quantitative). Asterisk code of significance: *significant at $P(0.05)$, **significant at $P(0.01)$, ***significant at $P(0.001)$.

	Residual mean square			F-ratio		
	Information Statistic	Indicator Species (qualitative)	Indicator Species (quantitative)	Information Statistic	Indicator Species (qualitative)	Indicator Species (quantitative)
soil pH	0.0464	0.0426	0.0429	7.43***	7.21***	8.90***
soil moisture	213.84	128.67	164.95	23.25***	31.18***	33.26***
height of ground	0.8335	0.4032	0.6376	41.50***	61.41***	57.47***
height of vegetation	604.00	677.38	661.93	3.81***	2.88**	2.56*
soil organic matter	449.59	300.36	407.10	8.97***	12.07***	11.00***
soil calcium	9.0927	12.195	11.078	3.41**	1.17	0.91
soil magnesium	1.5564	1.6897	1.8077	5.09***	2.56*	2.92*
soil potassium	10.809	7.5511	10.015	7.73***	6.97***	9.18***
soil sodium	1.0226	0.5326	0.7158	15.08***	28.94***	26.05***
soil manganese	0.0024	0.0013	0.0024	3.75**	5.59***	3.86**
soil iron	0.0299	0.0440	0.0350	3.69**	1.22	1.61
soil phosphorus	0.0288	0.0316	0.0224	1.90	2.08	5.41***
soil Mg : Ca ratio	0.0775	0.0553	0.0614	5.99***	5.54***	10.32***

yielding the most environmentally homogeneous associations. Because this method was also considered to have accorded best with the visual distribution of species assemblages on the ground, we may tentatively say that qualitative Indicator Species Analysis has given the best classification of stands on the Iping Common transect.

Detrended Correspondence Ordination (qualitative)

The stand and species results of DCO of the qualitative Iping Common data are shown in Figures 9.12 and 9.13, respectively. It is much more difficult to verbally describe the results of an ordination in broad outline. The graphical results resemble more closely the continuous nature of the vegetation than do the groupings produced by classification; but, while it is easy and profitable to describe the salient features of associations of stands or species, with ordinations it is more a question of picking out notable features.

A primary objective of ordination *sensu stricto* which, remember, may be regarded as Indirect Gradient Analysis, is to environmentally interpret the axes. Some investigations comprise vegetation data only; in this case, one can only attempt to interpret the axes environmentally by examining the species ordination and utilizing prior knowledge of the ecology of as many species for which such information is available. In the present instance, we

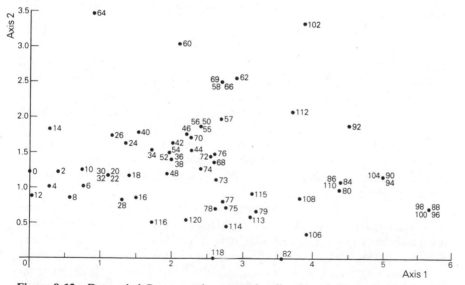

Figure 9.12 Detrended Correspondence stand ordination of the qualitative data of the Iping Common transect.

271

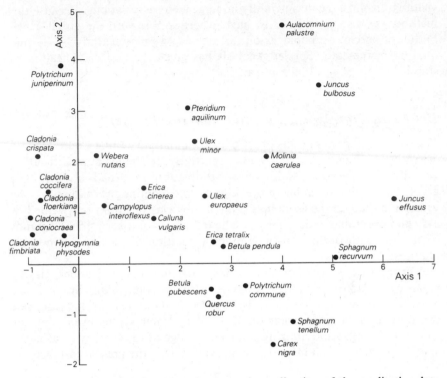

Figure 9.13 Detrended Correspondence species ordination of the qualitative data of the Iping Common transect.

find that wet heath species occur with high scores on Axis 1 (Fig. 9.13); and working towards the left, we encounter successively the species *Molinia caerulea*, *Erica tetralix*, *Calluna vulgaris*, and *Erica cinerea*. This sequence strongly suggests that Axis 1 is a soil moisture gradient. Axis 2 is very much more difficult, particularly on realizing that the four species having the highest and lowest scores on this axis occurred in only one or two stands: *Aulacomium palustre* (1), *Polytrichum juniperinum* (2), *Sphagnum tenellum* (1), *Carex nigra* (common sedge) (1). It is really impossible to make an intelligent guess as to the nature of Axis 2.

Using the available environmental data, however, gives a much more revealing picture of the ordination. Spearman's rank correlation coefficients are given in Table 9.11. The first axis is confirmed as a soil moisture gradient, having a high correlation coefficient of 0.67; but this axis is much more than simply a soil moisture gradient. Indeed, most of the environmental factors measured have highly significant correlations with the first axis through its stand scores; in particular, height of ground has the

Table 9.11 Spearman's rank correlation coefficients of qualitative Detrended Correspondence Ordination axis scores and environmental factors for the Iping Common transect. Asterisk code of significance levels as Table 9.10.

	Axis			
	1	2	3	4
soil pH	-0.66^{***}	0.16	0.32^{**}	-0.23
soil moisture	0.67^{***}	0.12	-0.56^{***}	0.45^{**}
height of ground	-0.94^{***}	0.28^{*}	0.48^{***}	-0.27^{*}
height of vegetation	0.51^{***}	0.05	-0.59^{***}	0.08
soil organic matter	0.51^{***}	0.11	-0.39^{**}	0.43^{***}
soil calcium	0.01	0.41^{***}	-0.17	0.39^{***}
soil magnesium	0.32^{**}	0.30^{**}	-0.30^{**}	0.45^{***}
soil potassium	0.47^{***}	0.21	-0.42^{***}	0.43^{***}
soil sodium	0.68^{***}	0.12	-0.52^{***}	0.31^{**}
soil manganese	0.48^{***}	0.15	-0.44^{***}	0.43^{***}
soil iron	0.45^{***}	0.16	-0.44^{***}	0.32^{**}
soil phosphorus	0.24^{*}	0.26^{*}	-0.31^{**}	0.25^{*}
soil Mg : Ca ratio	0.55^{***}	0.14	-0.45^{***}	0.32^{**}

unusually high coefficient for this kind of analysis of -0.94. Then comes soil sodium (0.68), soil moisture (0.67), and soil pH (-0.66). In fact, most of the soil nutrients have significant correlations with the first axis, but are mostly much lower in value; the notable exceptions are calcium and phosphorus – arguably the two most significant elements in plant nutrition affecting species distributions. So apart from calcium and magnesium, the first axis represents a complex gradient of soil nutrients, moisture, and organic matter, and soil pH and height of ground in the opposite direction.

The only factor to show a highly significant correlation on Axis 2 is soil calcium, but it is not very high and can only denote a very general trend. Indeed, inspection of the calcium levels of the stands with highest scores on this axis (Figs 9.12 & 4.2a) show them to be not especially high in this element.

The remaining two axes seem, on the whole, to be merely reflections of the first axis. There are probably no more than one, or possibly two, real gradients in this simple vegetation structure.

Detrended Correspondence Ordination (quantitative)

The stand and species results of DCO of the quantitative Iping Common data are shown in Figures 9.14 and 9.15, respectively; and the Spearman's correlation coefficients are given in Table 9.12. Generally, Axis 1 is very

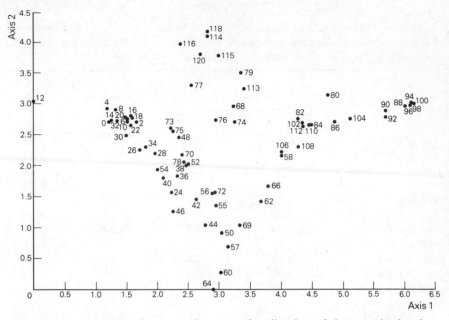

Figure 9.14 Detrended Correspondence stand ordination of the quantitative data of the Iping Common transect.

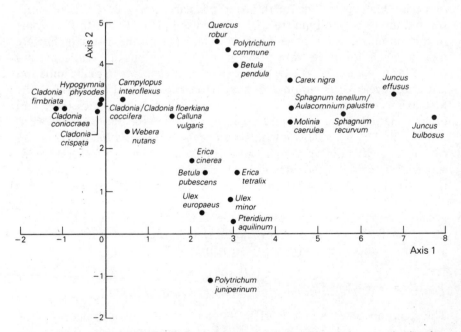

Figure 9.15 Detrended Correspondence species ordination of the quantitative data of the Iping Common transect.

274

Table 9.12 Spearman's rank correlation coefficients of quantitative Detrended Correspondence Ordination axis scores and environmental factors for the Iping Common transect. Asterisk code of significance levels as Table 9.10.

	Axis			
	1	2	3	4
soil pH	-0.68^{***}	-0.26	-0.15	0.18
soil moisture	0.68^{***}	-0.19	0.48^{***}	-0.19
height of ground	-0.90^{***}	-0.31^{*}	-0.31^{**}	0.14
height of vegetation	0.53^{***}	-0.12	0.34^{**}	-0.15
soil organic matter	0.50^{***}	-0.22	0.47^{***}	-0.08
soil calcium	0.03	-0.49^{***}	0.22	-0.09
soil magnesium	0.30^{**}	-0.41^{***}	0.40^{***}	-0.04
soil potassium	0.46^{***}	-0.35^{**}	0.45^{***}	-0.02
soil sodium	0.68^{***}	-0.14	0.39^{**}	-0.20
soil manganese	0.52^{***}	-0.26^{*}	0.23	-0.17
soil iron	0.44^{***}	-0.26^{*}	0.15	-0.13
soil phosphorus	0.28^{*}	-0.36^{**}	0.04	-0.06
soil Mg : Ca ratio	0.54^{***}	-0.28^{*}	0.41^{***}	-0.05

similar in respect of both qualitative and quantitative data, but differences become apparent on the second axis. In the quantitative data the correlations of the environmental factors with Axis 2 scores are all higher than in the qualitative data, with the important factors of soil magnesium, phosphorus, and potassium being particularly noticeable in this respect. Compare Axis 2 between Figures 9.12, 9.14, and between Figures 9.13 & 9.15, bearing in mind that there is a reversal in Axis 2 directions between the two analyses. We can thus infer that species such as *Polytrichum juniperinum*, *Pteridium aquilinum*, and *Ulex* spp. tend to be associated with the higher levels of soil calcium, magnesium, phosphorus, and potassium found in this transect; and *Polytrichum commune* and *Carex nigra* have opposite tendencies. Remember, however, that most of the species at the low nutrient end of Axis 2 have low frequencies, and so conclusions drawn about them may be less reliable. However, the difference between the two *Polytrichum* species is interesting in view of the tendency of *Polytrichum juniperinum* to occur early in the succession on recently burnt heathland (Watson 1968) where nutrient levels are likely to be higher.

Bray & Curtis Ordination (qualitative)

Although the use of a Bray & Curtis Ordination is pointless if facilities exist for computing the Detrended Correspondence Ordination, the former method is presented here in order to highlight the problems involved even with a rather small data set. First, the steps involved in its construction are given. The formulation of the method allows several different pathways to a final ordination, and both ecological knowledge and subjective judgements are necessary to produce the final product.

With 70 stands in the Iping Common transect, there are $(70)(69)/2 = 2415$ similarity values. Of these, 725 have the maximum value of 1. How can we select one of these whose stands will then define the ends of the first axis? From what has been described already, both here and in Chapter 4, it is clear that the main vegetation changes and many of the environmental factors are correlated with height of ground. Accordingly, we choose as ends of the first axis Stand 0 m (the highest of the transect) and Stand 100 m (the lowest). With hindsight, these are also the stands at the extreme ends of the first axis of the DCO (Fig. 9.12). Designating the score of Stand 0 m as 0 and that of Stand 100 m as 1, the scores of the remaining stands are calculated as shown in Chapter 7.

It could also be argued that soil moisture is the most important environmental gradient on this transect, and that the stands with the highest and lowest soil moistures should have been used to define the first axis. However, soil moisture and height of ground are fairly highly correlated (-0.5926). Further, it may be argued that height of ground is a better indicator of *long-term* soil moisture conditions than is the instantaneous soil moisture content actually measured (and this during quite a severe drought).

Returning to the ordination, 12 stands have scores of 0.5 on Axis 1, and among these stands there are 66 dissimilarity values. Ten pairs of stands have the maximum dissimilarity value of 1, between Stands 60 and 64 m on the one hand, and Stands 79, 82, 102, 112, and 113 m on the other. The former pair are in the dense bracken zone, while the latter group have *Molinia caerulea* constantly present along with a moss, usually *Polytrichum commune*. Stand 112 m, however, contained only *Molinia caerulea*, and Stand 60 m comprised only *Pteridium aquilinum*. When these stands were used as endpoints of Axis 2, the resulting stand scores on this axis were mostly 0.5, while the remainder were concentrated at a few other scores: the amount of information furnished by Axis 2 was small. It is necessary to use species-rich stands as endpoints to obtain a variety of stand scores along an axis; accordingly, Stand 64 m containing two species and Stand 113 m containing four species were selected. This choice of endpoint stands gave a much better ordination for the first two axes, and is shown in Figure 9.16.

At this stage there are still many groups, shown ringed in Figure 9.16, whose member stands have identical scores on the first two axes. Some of these groups are wholly or partially composed of stands having identical species content (dissimilarity zero); however, some stands within some groups are not identical, but merely project to the same scores on Axes 1

276

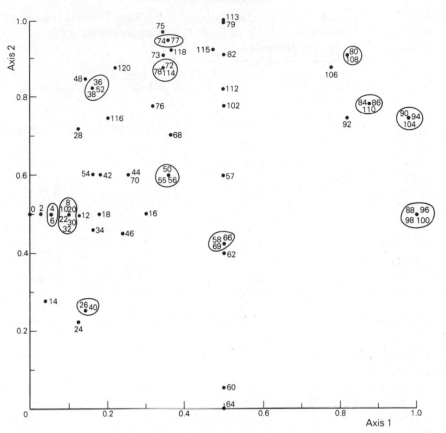

Figure 9.16 Bray & Curtis stand ordination of the qualitative data of the Iping Common transect.

and 2. The maximum dissimilarity of a pair of stands within any one group is 0.67 (i.e. Stands 8 & 10 m and 80 & 108 m); but since there are many pairs of stands remaining, which are not already endpoints of axes, with dissimilarity values of 1, it is not possible to use a pair of stands within any of the clusters as the endpoints of Axis 3 without introducing distortion into this axis. Accordingly, we look for two stands which: (a) are near the middle of the plane formed by the first two axes; (b) are close together; (c) have a dissimilarity of unity. Stands 16 & 50 m are the only obvious candidates, and so are placed at the ends of Axis 3. When the scores of the other 68 stands on Axis 3 have been calculated it is impossible to find two further stands which are close together in respect of their scores on the first three axes and have a dissimilarity of 1. Thus we terminate the procedure here, and do not define a fourth axis. Even so, we have defined one extra

277

Table 9.13 Spearman's rank correlation coefficients of Bray & Curtis Ordination axis scores and environmental factors for the Iping Common transect (qualitative data). Asterisk code of significance levels as Table 9.10.

	Axis		
	1	2	3
soil pH	-0.68^{***}	-0.31^{**}	-0.24
soil moisture	0.68^{***}	0.17	0.53^{***}
height of ground	-0.90^{***}	-0.51^{***}	-0.39^{***}
height of vegetation	0.52^{***}	0.06	0.40^{***}
soil organic matter	0.50^{***}	0.19	0.47^{***}
soil calcium	0.01	-0.14	0.30^{*}
soil magnesium	0.28^{*}	0.03	0.42^{***}
soil potassium	0.44^{***}	0.14	0.47^{***}
soil sodium	0.66^{***}	0.21	0.53^{***}
soil manganese	0.53^{***}	0.26^{*}	0.56^{***}
soil iron	0.43^{***}	0.25^{*}	0.45^{***}
soil phosphorus	0.28^{*}	0.04	0.42^{***}
soil Mg : Ca ratio	0.53^{***}	0.16	0.48^{***}

potential gradient over and above the two that were deemed to exist as a result of DCO.

The Spearman's rank correlation coefficients of environmental factors with Bray & Curtis Ordination axis scores are shown in Table 9.13. The interpretation of Axis 1 is identical with that of the two previous ordinations (Tables 9.5 & 9.6), but the interpretations of the other axes differ from one ordination to another. In particular the Bray & Curtis Axis 2 seems rather meaningless in terms of the environmental factors measured in this study.

Clearly, the Bray & Curtis Ordination applied to the Iping Common data has reproduced the very obvious gradient on the first axis. Any other gradient which might have existed in the vegetation of the data set is very vague: DCO gives some indication of a definite secondary gradient, but Bray & Curtis Ordination does not. The species ordination, shown in Figure 9.17, is not strictly a Bray & Curtis Ordination, but is constructed by the centroid method (p. 239) based on the Bray & Curtis stand ordination.

Classifications with ordinations

A useful graphical procedure, which helps in the further direct interpretation of an ordination diagram and also serves to unify the two broad approaches, is to delimit areas on an ordination diagram representing

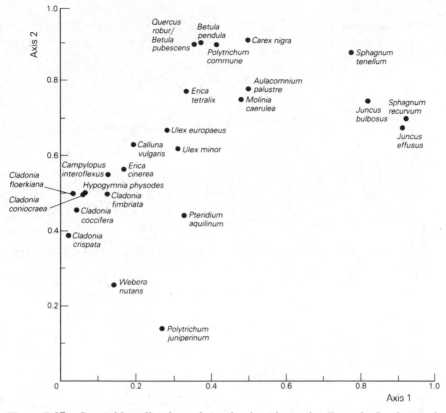

Figure 9.17 Centroid ordination of species based on the Bray & Curtis stand ordination of the qualitative data of the Iping Common transect.

particular associations obtained from a classification procedure. Figure 9.18 shows two examples, both using DCO of qualitative data: (a) Information Statistic Analysis; (b) Indicator Species Analysis of qualitative data. Each diagram is constructed simply by enclosing areas containing points representing stands belonging to the same association. Sometimes the picture is clear cut, as in Figure 9.18b, but on other occasions there is introgression of one association into another, as with Association E into B in Figure 9.18a. Another feature of Figure 9.18a is the very scattered nature of the stands of Association G. Consequently, the best way of representing this association is simply as being outside all the other associations. This highlights the nature of the final association in a divisive monothetic classification, defined by absences of all divisor species, in that it is often a motley collection of unrelated stands.

In Figure 9.18b, Stand 102 m is clearly rather different from the other

279

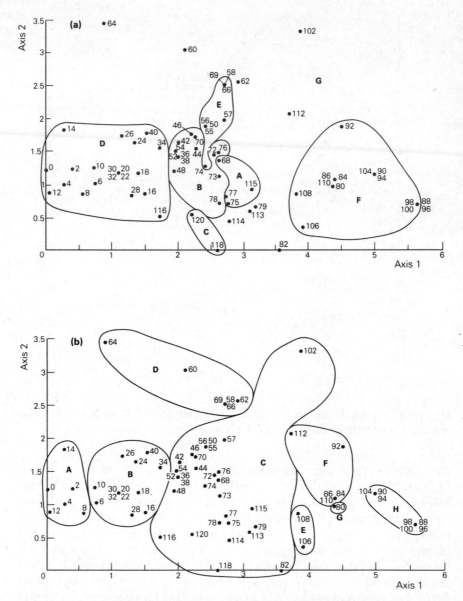

Figure 9.18 Detrended Correspondence Ordination of the Iping Common transect with stands grouped by associations produced by: (a) Information Statistic classification; (b) Indicator Species Analysis.

stands in its association, C. The same classification procedure on a data matrix from which species whose frequencies are only 1, or 1 & 2, are eliminated (MINO = 2 or 3) moves Stand 102 m to Association F. Thus the more 'mobile' nature of this stand with regard to its classification is reflected in its ordination position relative to other stands.

Direct Gradient Analysis – one factor

A useful species ordination is obtained by a Direct Gradient Analysis on the soil moisture gradient – obviously a most important environmental factor on this transect (Fig. 9.19). Little comment is required, other than to question the reality of the dip in the mid ranges of the *Calluna vulgaris* and *Molinia caerulea* curves. They do appear to be mirror images of one another, which might be expected of two species which attain dominance in suitable habitats; but further detailed work would be required to elucidate this phenomenon.

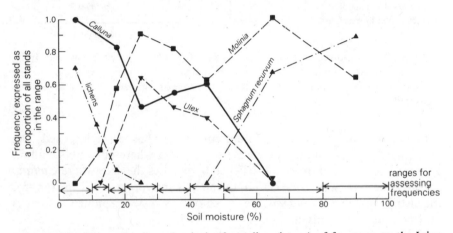

Figure 9.19 Direct Gradient Analysis (for soil moisture) of five taxa on the Iping Common transect.

Semi-direct Gradient Analysis

Two examples are shown of Semi-direct Gradient Analyses. In each case the variance–covariance matrix of the environmental factors has been submitted to Principal Component Analysis, the results of which are given in Table 7.2, and the first two axes of the stand ordination are shown in Figure 7.11a. Figure 9.20 shows the distribution of *Calluna vulgaris*. Each rectangular cell shows the proportion of stands within the area of the cell that contains *Calluna vulgaris*, cells without a number have no stands

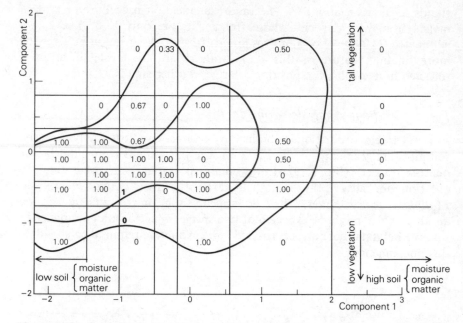

Figure 9.20 Semi-direct Gradient Analysis of *Calluna vulgaris* on the Iping Common transect line, based on the Principal Component Analysis of the variance–covariance matrix of a sub-set of five of the environmental factors measured.

within their area. Finally, two contour curves have been superimposed. The inner, labelled 1, approximately encloses an area where all stands contain *Calluna vulgaris*; and the outer, labelled 0, excloses the area where *Calluna vulgaris* does not occur. In relation to the axis interpretations (pp. 163–5), the distribution of *Calluna vulgaris* in relation to key environmental factors may be quickly appraised.

In Figure 9.21 the Semi-direct Gradient Analysis of two taxa is given: *Sphagnum recurvum*, and all the lichen species taken together. Only a single contour curve is given for each taxon, delimiting the area of presence from the area of absence. It is interesting to note the outliers, particularly of *Sphagnum recurvum*.

Figure 9.22 shows a more conventional species ordination of the Semi-direct Gradient Analysis type. Here, the correlation matrix of the environmental factors has been submitted to Principal Component Analysis (Table 7.3), and the position of each species is given by the centroid of its occurrence in the stands (Ch. 8, p. 239). The direction of Axis 1 is reversed in comparison with the situation in Figures 9.20 and 9.21. This species ordination should be compared with the other species ordinations shown

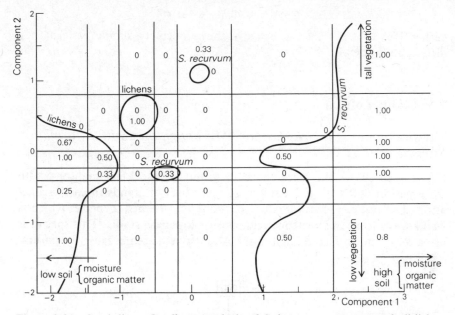

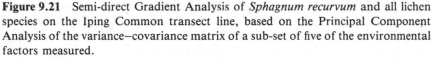

Figure 9.21 Semi-direct Gradient Analysis of *Sphagnum recurvum* and all lichen species on the Iping Common transect line, based on the Principal Component Analysis of the variance–covariance matrix of a sub-set of five of the environmental factors measured.

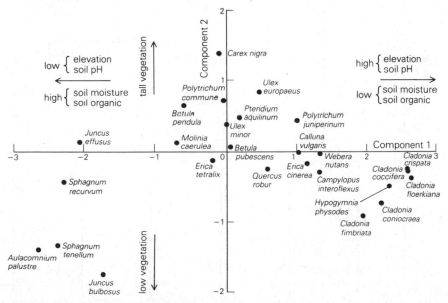

Figure 9.22 Centroid ordination of all species on the Iping Common transect line, based on the Principal Component Analysis of the correlation matrix of a sub-set of five of the environmental factors measured.

283

earlier (Figs 9.13, 9.15, 9.17). Broadly, they all depict a similar pattern; they differ only in detail.

Coed Nant Lolwyn

Divisive monothetic classification: Macnaughton-Smith Information Statistic (normal)

Figure 9.23 shows the hierarchical diagram for the Macnaughton-Smith Information Statistic, and Table 9.14 gives lists of stands in each association. The primary division is on *Mercurialis perennis*, a species which is well known to be absent from acid soils of low base status. This species is not a common woodland plant in the Aberystwyth area because the natural

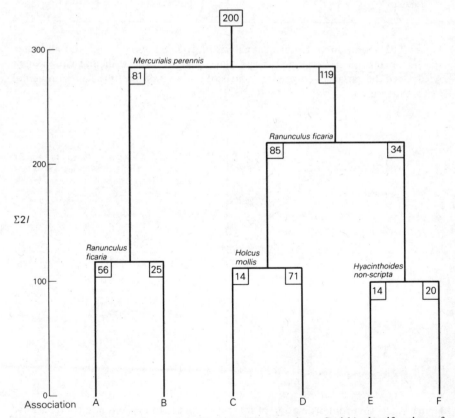

Figure 9.23 Normal Information Statistic (Macnaughton-Smith) classification of the Coed Nant Lolwyn study.

Table 9.14 List of stands in each association of the normal Information Statistic of the Coed Nant Lolwyn data.

Association	Number of stands	List of stands
A	56	10, 15, 20, 27, 28, 32, 33, 37, 54, 55, 59, 61, 69, 71, 74, 76, 78, 82, 87, 88, 89, 90, 91, 95, 96, 98, 102, 103, 105, 106, 107, 108, 110, 111, 112, 113, 114, 118, 121, 122, 126, 128, 129, 135, 136, 137, 138, 139, 141, 143, 150, 157, 159, 169, 197, 199
B	25	22, 43, 44, 45, 58, 60, 72, 75, 79, 92, 93, 94, 101, 104, 109, 119, 124, 125, 130, 131, 134, 140, 145, 146, 173
C	14	1, 3, 6, 7, 11, 12, 16, 35, 62, 65, 66, 77, 181, 193
D	71	2, 4, 8, 14, 19, 21, 29, 31, 34, 36, 39, 40, 46, 50, 53, 64, 67, 68, 70, 73, 81, 84, 86, 97, 99, 100, 115, 116, 117, 120, 123, 132, 133, 142, 144, 147, 148, 149, 151, 152, 153, 154, 155, 156, 158, 160, 161, 162, 163, 164, 165, 166, 167, 168, 171, 172, 174, 175, 176, 177, 178, 179, 182, 183, 187, 188, 189, 190, 191, 198, 200
E	14	26, 38, 41, 42, 47, 48, 49, 51, 52, 56, 57, 63, 83, 192
F	20	5, 9, 13, 17, 18, 23, 24, 25, 30, 80, 85, 127, 170, 180, 184, 185, 186, 194, 195, 196

soils of the district tend to be acid and nutrient poor. Hence the occurrence of *Mercurialis perennis* as the first divisor species is ecologically significant, separating the base-rich areas of the wood from the more acidic, base-poor areas. Figure 9.24a,b,d,f shows this effect, where Associations A and B should be compared with Associations C to F taken together. Further, *Mercurialis perennis* is intolerant of waterlogged soils (Martin 1968), a condition indicated by a high extractable manganese content associated with the element in the divalent form. It is noteworthy that Associations A and B, particularly A, have low manganese contents (Fig. 9.24e).

There seems little reason for splitting the stands containing *Mercurialis perennis* into the two associations A and B, on the basis of presence or absence of *Ranunculus ficaria* (lesser celendine). The only environmental factor to differ markedly between the two associations is soil extractable potassium (Fig. 9.24c); but, on average, Association B has slightly lower soil pH and magnesium, and higher soil manganese, and vernal and aestival

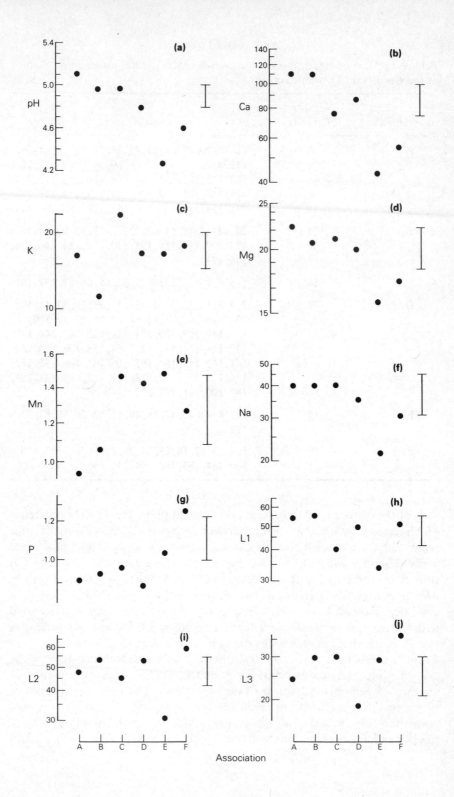

light levels (Fig. 9.24a,d,e,i,j). Besides the necessary difference in *Ranunculus ficaria*; *Anemone nemorosa*, *Geum urbanum* (wood avens), *Circaea lutetiana*, *Geranium robertianum* (herb Robert) and *Sanicula europaea* are markedly less frequent in Association B than in A (Figs 9.25e; 9.26e,f,g; 9.27e; 9.28a); while *Oxalis acetosella*, *Silene dioica* (red campion) and *Urtica dioica* (stinging nettle) (Figs 9.27c; 9.28c,e), together with several bryophytes (*Atrichum undulatum*, *Brachythecium rutabulum*, *Eurhynchium praelongum*, *Eurhynchium striatum*, *Plagiochila asplenioides*, *Plagiomnium undulatum* and *Thuidium tamariscinum*) are notably more frequent in B than in A. The mean number of species per stand is not markedly different (9.5 in Association A, 8.6 in Association B), but the species richness (number of species in the stands of an association) is much less in Association B than in A (Table 9.15).

Although there was some indication of a higher base content in soils of stands containing *Ranunculus ficaria* in the presence of *Mercurialis perennis*, this difference becomes more pronounced in the absence of *Mercurialis perennis* − Associations C and D contrasted with E and F (Figs 9.24a,b,d,f). Again, the pattern of variation of potassium differs from that of pH, calcium, magnesium and sodium; and also, as before, the main difference between Associations C and D is one of potassium levels (Association C has the highest mean potassium level out of all the groups), but this time associated with a marked difference in aestival light levels (Fig. 9.24c,j). With regard to the major species differences between Associations C and D, *Galium aparine* (goosegrass), *Geum urbanum*, *Circaea lutetiana*, *Veronica montana*, and *Urtica dioica* are markedly less frequent in Association C than in D (Figs 9.26c,e,g; 9.27a; 9.28c) with *Ranunculus ficaria* having a significantly lower density in the former group (Fig. 9.25d); while many other species have higher frequencies in Group C than in D, particularly the grasses (Table 9.15).

In contrast, Associations E and F differ markedly, but not necessarily significantly in the statistical sense, in several environmental factors − soil pH, calcium, magnesium, manganese, sodium, phosphorus, and pre-vernal and vernal light regimes. Apart from the absolute presence or absence of *Hyacinthoides non-scripta* (the divisor species), *Anemone nemorosa* and *Oxalis acetosella* (Figs 9.25e; 9.27c) together with *Stellaria holostea* (greater stitchwort), *Eurhynchium striatum*, *Plagiochila asplenioides*, and *Thuidium*

Figure 9.24 Mean values of environmental factors in the normal Information Statistic associations of the Coed Nant Lolwyn study: (a) soil pH; soil extractable (b) calcium, (c) potassium, (d) magnesium, (e) manganese, (f) sodium, (g) phosphorus; light intensities expressed as a percentage of full daylight (h) pre-vernal, (i) vernal, (j) aestival. The approximate least significant ranges are at $P(0.05)$. Units of nutrient element concentrations are mg $(100\,g)^{-1}$ dry soil.

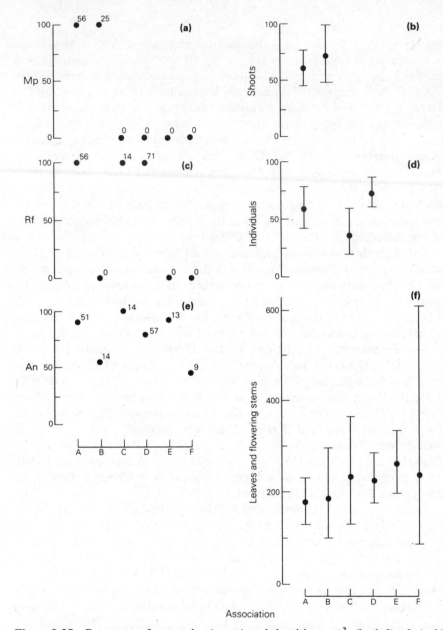

Figure 9.25 Percentage frequencies (a, c, e) and densities, m⁻², (b, d, f) of: (a, b) *Mercurialis perennis*, (c, d) *Ranunculus ficaria*, (e, f) *Anemone nemorosa*; in associations produced by Information Statistic classification in the Coed Nant Lolwyn study. The numbers adjacent to the points on the frequency graphs give the actual number of stands within each association in which the species occurs; and the lengths of the vertical lines attached to the points on the density graphs equal one standard error above and below the mean, evaluated on square root transformed data.

288

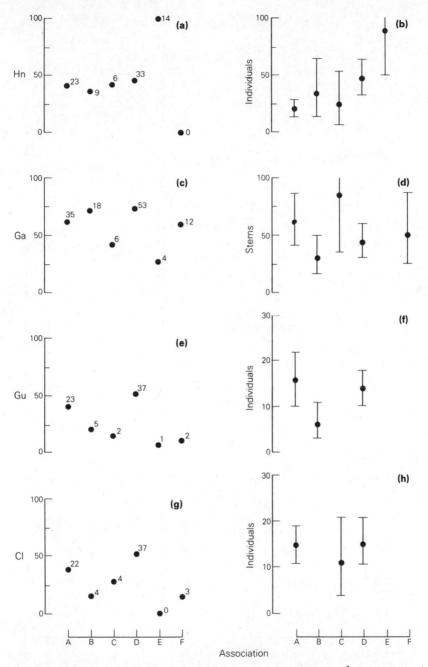

Figure 9.26 Percentage frequencies (a, c, e, g) and densities, m^{-2}, (b, d, f, h) of: (a, b) *Hyacinthoides non-scripta*, (c, d) *Galium aparine*, (e, f) *Geum urbanum*, (g, h) *Circaea lutetiana*; in associations produced by Information Statistic classification in the Coed Nant Lolwyn study. Other details as Figure 9.25.

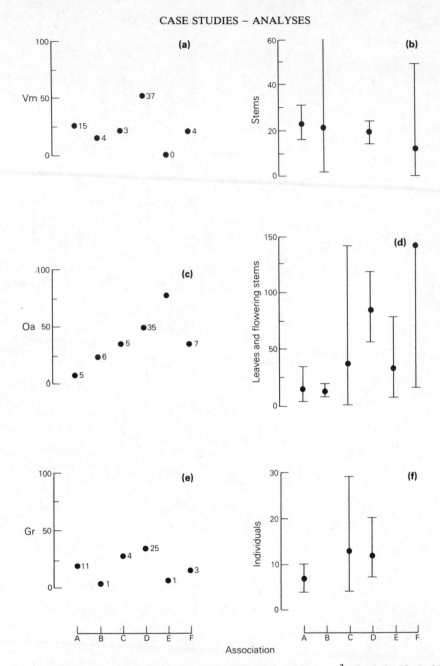

Figure 9.27 Percentage frequencies (a, c, e) and densities, m^{-2}, (b, d, f) of: (a, b) *Veronica montana*, (c, d) *Oxalis acetosella*, (e, f) *Geranium robertianum*; in associations produced by Information Statistic classification in the Coed Nant Lolwyn study. Other details as Figure 9.25.

290

Table 9.15 Coed Nant Lolwyn, normal Information Statistic Analysis. The percentage occurrence of each species within each association.

	Association					
	A	B	C	D	E	F
Agrostis capillaris	0	0	7	1	0	5
Anthoxanthum odoratum	0	0	7	1	0	10
Arrhenatherum elatius	1	4	7	2	0	0
Brachypodium sylvaticum	1	0	28	11	14	0
Dactylis glomerata	1	0	21	5	0	10
Festuca gigantea	0	0	14	0	0	5
Holcus mollis	5	4	100	0	50	35
Luzula campestris	0	0	7	0	0	15
Melica uniflora	7	12	28	14	21	5
Milium effusum	1	4	0	15	7	5
Poa nemoralis	7	4	7	7	0	5
Poa trivialis	5	0	7	14	7	0
Adoxa moschatellina	5	4	7	18	7	20
Anemone nemorosa	91	56	100	80	92	45
Chrysosplenium oppositifolium	5	16	0	7	0	5
Circaea lutetiana	39	16	28	52	0	15
Conopodium majus	3	0	28	8	7	0
Epilobium montanum	0	0	0	7	0	10
Filipendula ulmaria	14	0	28	4	0	5
Fragaria vesca	0	0	7	7	0	0
Galium aparine	62	72	42	74	28	60
Geranium robertianum	19	4	28	35	7	15
Geum urbanum	41	20	14	52	7	10
Glechoma hederacea	5	8	0	2	7	10
Heracleum sphondylium	7	4	28	7	0	0
Hyacinthoides non-scripta	41	36	42	46	100	0
Lapsana communis	0	0	0	4	0	5
Mercurialis perennis	100	100	0	0	0	0
Moehringia trinervia	0	8	7	1	7	15
Oxalis acetosella	8	24	35	49	78	35
Potentilla sterilis	5	0	7	16	0	0
Primula vulgaris	1	0	42	8	0	10
Ranunculus ficaria	100	0	100	100	0	0
Rumex crispus	0	0	0	2	0	5
Sanicula europaea	12	0	28	29	7	5
Silene dioica	3	16	21	14	14	55
Solidago virgaurea	0	0	14	0	0	10
Stachys betonica	0	0	14	0	14	0
Stachys sylvatica	3	0	0	2	0	5

Table 9.15 (*continued*)

	A	B	C	D	E	F
			Association			
Stellaria holostea	10	0	57	32	35	0
Taraxacum officinale	1	0	21	12	0	0
Urtica dioica	10	32	0	14	7	35
Veronica chamaedrys	7	0	57	19	7	5
Veronica hederifolia	12	0	0	2	0	5
Veronica montana	26	16	21	52	0	20
Viola riviniana	1	4	35	15	14	10
Acer pseudoplatanus	8	8	14	5	7	10
Crataegus monogyna	3	0	7	4	0	5
Fraxinus excelsior	7	0	28	21	7	5
Hedera helix	16	24	35	30	42	35
Lonicera periclymenum	0	0	7	4	7	5
Prunus spinosa	3	0	0	8	0	0
Rosa canina	1	4	14	5	0	0
Rubus fruticosus	10	16	14	21	0	20
Rubus idaeus	0	0	0	4	0	5
Atrichum undulatum	1	12	7	7	0	5
Brachythecium rutabulum	5	16	7	5	0	15
Brachythecium velutinum	5	0	0	5	0	0
Cirriphyllum piliferum	3	0	0	2	0	10
Eurhynchium praelongum	58	76	57	73	64	60
Eurhynchium stratium	39	64	35	42	85	25
Fissidens bryoides	3	0	7	1	0	0
Fissidens taxifolius	7	0	0	0	0	0
Lophocolea bidentata	5	8	0	5	14	10
Mnium hornum	3	4	0	14	28	10
Plagiochila asplenioides	8	28	14	11	42	15
Plagiomnium undulatum	28	48	28	23	7	10
Plagiothecium denticulatum	0	4	0	4	14	0
Plagiothecium nemorale	5	8	0	5	28	10
Polytrichum formosum	1	0	0	0	14	5
Rhynchostegium confertum	10	4	14	4	0	10
Rhytidiadelphus loreus	0	0	0	2	7	5
Thamnobryum alopecurum	3	4	7	1	0	0
Thuidium tamariscinum	10	36	14	11	50	15
Athyrium filix-femina	5	0	0	0	0	0
Dryopteris dilatata	3	16	7	9	14	5
Dryopteris filix-mas	0	0	0	5	21	0

tamariscinum are notably more frequent in E than in F; while *Galium aparine*, *Silene dioica*, and *Urtica dioica* are more frequent in the latter association (Figs 9.26c; 9.28c,e).

The main features of the associations may be summarized as follows. Stands of Associations A and B occur on relatively base-rich soils of low extractable manganese status and high pre-vernal light, and all the stands contain *Mercurialis perennis*. Relative to A, Association B contains many fewer species, does not contain *Ranunculus ficaria*, but does have higher frequencies of several bryophyte species. Association B has a very low mean soil potassium level. Both associations have essentially similar densities of *Mercurialis perennis*, and also of *Oxalis acetosella* which are very low in comparison with the other associations (Figs 9.25b; 9.27d).

Stands of Association C tend to have high soil potassium and manganese levels, a moderate base status, and a rather low pre-vernal light level. The vegetation can be described as grassy (apart from the total absence of the woodland grass *Milium effusum*). Besides the grass species, *Stellaria holostea* and *Veronica chamaedrys* (germander speedwell), which are typical of semi-shade hedgerow grassland, are frequent in this association.

Association D is the largest of the six, and is distinguished environmentally by low soil phosphorus and low aestival light levels (Figs 9.24g,j). It can best be described as a typical woodland community on moderately acid soil. Apart from the ubiquitous *Ranunculus ficaria*, which has quite a high density (Fig. 9.25d); *Anemone nemorosa*, *Galium aparine*, and *Eurhynchium praelongum* have a frequency of 74% or more, and *Circaea lutetiana*, *Geum urbanum*, and *Veronica montana* have just over 50% frequency (Table 9.15). *Hyacinthoides non-scripta* and *Oxalis acetosella* have just under 50% frequency, but show a high density (Figs 9.26a,b; 9.27c,d).

The remaining two associations differ considerably from one another. Association E has the lowest soil pH and base status, and the lowest pre-vernal light levels. It is species poor, having *Hyacinthoides non-scripta* as the constant species, together with *Anemone nemorosa*, *Eurhynchium striatum*, and *Oxalis acetosella* having frequencies in excess of 75%. *Eurhynchium praelongum*, *Holcus mollis*, and *Thuidium tamariscinum* are present in at least half the stands (Table 9.15).

In contrast, Association F is characterized environmentally by high soil phosphorus and high vernal and aestival light levels (Figs 9.24g,i,j). The stands in this association are mostly near the long south-eastern boundary with agricultural land, which explains the high phosphorus and light levels found. Species frequencies are generally low, the highest being *Eurhynchium praelongum* and *Galium aparine* at 60%, followed by *Silene dioica* (55%), *Anemone nemorosa* (45%), and *Holcus mollis*, *Oxalis acetosella*, *Urtica dioica*, and *Hedera helix* all at 35% (Table 9.15).

293

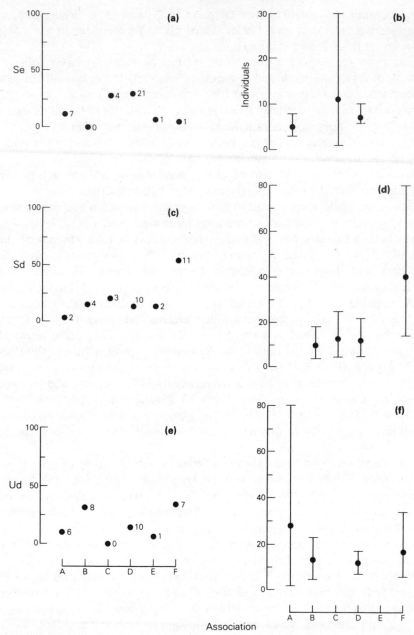

Figure 9.28 Percentage frequencies (a, c, e) and densities, m^{-2}, (b, d, f) of: (a, b) *Sanicula europaea*, (c, d) *Silene dioica*, (e, f) *Urtica dioica*; in associations produced by Information Statistic classification in the Coed Nant Lolwyn study. Other details as Figure 9.25.

Divisive polythetic classification: Indicator Species Analysis (qualitative, normal)

The hierarchical diagram is shown in Figure 9.29. The first division is very uneven, splitting off just 22 of the 200 stands on the positive side. The reason for this can be appreciated by examining the distribution of stands along the first axis of the Detrended Correspondence Ordination (Fig. 9.37); there are only a few stands occupying a considerable length of the right-hand side of this axis. The only common environmental feature of the two associations, E and F, on the positive side is low soil calcium levels (Fig 9.30b), but in many other respects the two associations produced from these 22 stands are quite different.

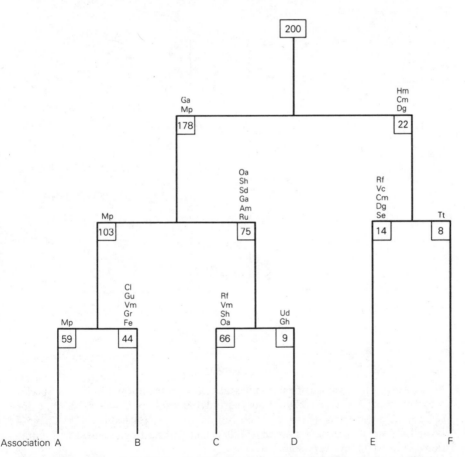

Figure 9.29 Normal Indicator Species Analysis (TWINSPAN) of the qualitative data of the Coed Nant Lolwyn study.

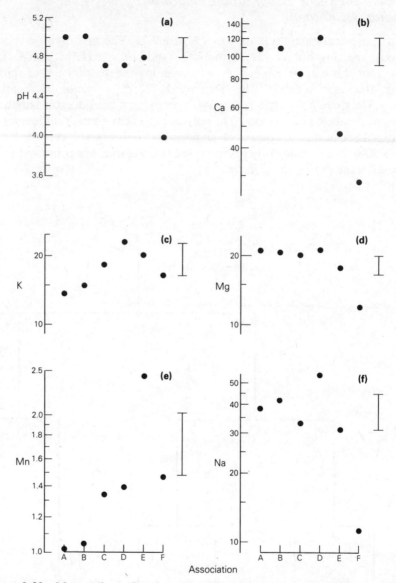

Figure 9.30 Mean values of environmental factors in the normal Indicator Species associations of the Coed Nant Lolwyn study: (a) soil pH; soil extractable (b) calcium, (c) potassium, (d) magnesium, (e) manganese, (f) sodium. The approximate least significant ranges are at $P(0.05)$. Units of nutrient element concentrations are mg $(100\ g)^{-1}$ dry soil.

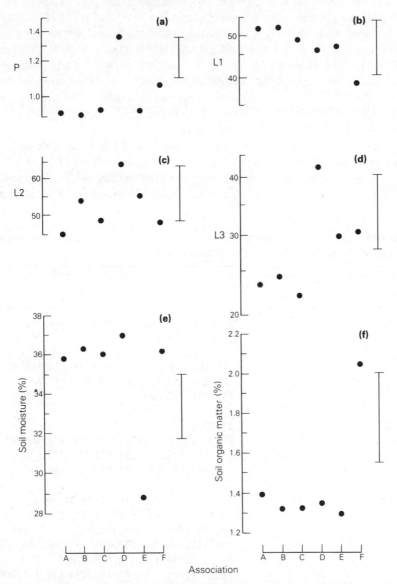

Figure 9.31 Mean values of environmental factors in the normal Indicator Species associations of the Coed Nant Lolwyn study: (a) soil extractable phosphorus mg $(100\,\mathrm{g})^{-1}$ dry soil light; intensities expressed as a percentage of full daylight (b) pre-vernal, (c) vernal, (d) aestival; (e) soil moisture, (f) soil organic matter. The approximate least significant ranges are at $P(0.05)$.

297

From the viewpoint of several environmental factors, Association F represents an extreme, having particularly low levels of soil pH, calcium, magnesium, and sodium, but outstandingly high soil organic matter, together with low pre-vernal and vernal light regimes (Figs 9.30a,b,d,f; 9.31b,c,f). *Holcus mollis* is the only constant species, while *Anemone nemorosa, Oxalis acetosella, Thuidium tamariscinum, Hyacinthoides non-scripta, Eurhynchium striatum*, and *Plagiochila asplenioides* are frequent (Table 9.17). Many of the common woodland species are absent, and the association is the most species poor of the set.

Association E is characterized environmentally by low soil calcium and moisture, together with a very high soil manganese level (Figs 9.30b,e; 9.31e). *Anemone nemorosa* is the only constant species, while *Ranunculus ficaria* and *Holcus mollis*, and to a lesser extent *Conopodium majus, Oxalis*

Table 9.16 List of stands in each association of the normal Indicator Species Analysis of the Coed Nant Lolwyn (qualitative) data.

Association	Number of stands	List of stands
A	59	2, 20, 28, 32, 33, 38, 44, 45, 47, 48, 54, 55, 59, 60, 72, 74, 76, 78, 82, 84, 85, 87, 88, 92, 94, 95, 101, 102, 104, 106, 108, 109, 110, 111, 112, 113, 114, 116, 118, 119, 121, 124, 125, 126, 127, 128, 129, 130, 131, 134, 136, 140, 141, 146, 150, 169, 173, 197, 199
B	44	12, 22, 27, 31, 35, 36, 37, 50, 53, 57, 58, 67, 69, 89, 90, 91, 93, 96, 97, 98, 99, 100, 103, 105, 107, 115, 117, 120, 122, 123, 132, 133, 135, 138, 142, 143, 144, 145, 148, 149, 157, 164, 165, 183
C	66	8, 14, 15, 16, 17, 18, 19, 21, 29, 34, 39, 40, 41, 43, 49, 62, 63, 64, 68, 70, 71, 73, 75, 77, 81, 83, 86, 137, 139, 147, 151, 152, 153, 154, 155, 156, 158, 159, 160, 161, 162, 163, 166, 167, 168, 170, 171, 172, 174, 175, 176, 177, 178, 179, 180, 181, 182, 184, 185, 186, 187, 188, 190, 191, 198, 200
D	9	13, 23, 24, 25, 26, 30, 61, 79, 80
E	14	1, 3, 4, 5, 6, 7, 9, 10, 11, 46, 65, 66, 189, 193
F	8	42, 51, 52, 56, 192, 194, 195, 196

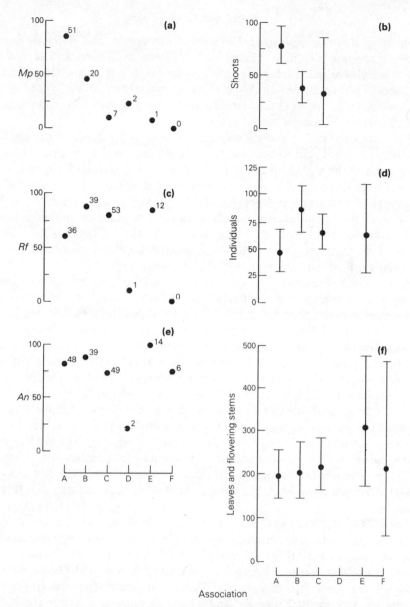

Figure 9.32 Percentage frequencies (a, c, e) and densities, m^{-2}, (b, d, f) of: (a, b) *Mercurialis perennis*, (c, d) *Ranunculus ficaria*, (e, f) *Anemone nemorosa*; in associations produced by Indicator Species Analysis in the Coed Nant Lolwyn study. The numbers adjacent to the points on the frequency graphs give the actual number of stands within each association in which the species occurs; and the lengths of the vertical lines attached to the points on the density graphs equal one standard error above and below the mean, evaluated on square root transformed data.

acetosella, Stellaria holostea, Veronica chamaedrys, and *Eurhynchium praelongum* are frequent (Table 9.17). *Geum urbanum* and *Circaea lutetiana*, absent in Association F, are present in E but at low frequencies and densities (Figs 9.33e–h). Association E represents a transition between the species assemblage of Association F on acid, base-poor soils and species assemblages occuring on base-rich soils.

The nine stands of Association D are not part of the gradient of species assemblages from acid to base-rich soils. Calcium, potassium and sodium are present at the highest mean levels in the soils of these stands; vernal and aestival light levels are also high, but the outstanding environmental feature is the high soil phosphorus level (Figs 9.30b,c,f; 9.31a,c,d). These stands all occur near the south-eastern boundary, already referred to in connection with Association F of the Information Statistic classification; but Association D of the present classification is more clearly defined than is Association F of the Information Statistic classification.

Associations A and B occur on base-rich soils of low manganese and phosphorus content (Figs 9.30a,b,d,e,f; 9.31a). There are no obvious differences between the two associations with respect to the environmental factors measured, but the species assemblages are quite different. In Association A the most frequent species is *Mercurialis perennis*, followed by *Anemone nemorosa, Galium aparine, Ranunculus ficaria*, and *Eurhynchium praelongum*. The last three species, however, have a frequency of only just over 60% (Table 9.17). The densities of *Mercurialis perennis* are generally high, and those of *Ranunculus ficaria* low (Figs 9.32b,d). *Oxalis acetosella* is present at very low frequency and density (Figs 9.34c,d). In contrast, Association B contains both a higher frequency and density of *Ranunculus ficaria*, and considerable higher frequencies of *Geum urbanum, Circaea lutetiana*, and *Veronica montana* (Figs 9.32c,d, 9.33e,g, 9.34a). *Mercurialis perennis* has much lower frequencies and densities (Figs 9.32a,b).

The soils of Association C are, on the whole, less base-rich than are those of the two previous associations; also, soil potassium and manganese levels are markedly higher (Figs 9.30a,b,c,e,f). *Galium aparine, Ranunculus ficaria, Eurhynchium praelongum, Anemone nemorosa* and *Oxalis acetosella* are the most frequent species (Table 9.9), with *Ranunculus ficaria* having an intermediate density and *Oxalis acetosella* a relatively high density (Figs 9.32d, 9.34d).

It seems as though the association sequence A, B, C, E, F represents a gradient of increasing soil acidity and declining base content (particularly calcium). Further, the aestival light levels of Associations A, B, C are markedly lower than those of Associations E and F (Fig. 9.31d). Association D represents a more phosphate-rich, wood edge environment which does not fit into the soil base status gradient of the other associations.

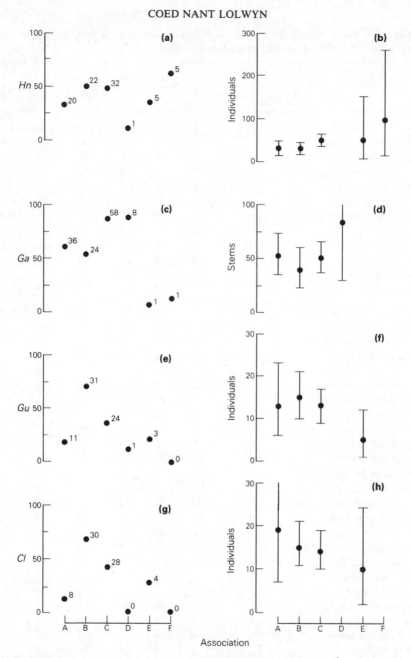

Figure 9.33 Percentage frequencies (a, c, e, g) and densities, m^{-2}, (b, d, f, h) of: (a, b) *Hyacinthoides non-scripta*, (c, d) *Galium aparine*, (e, f) *Geum urbanum*, (g, h) *Circaea lutetiana*; in associations produced by Indicator Species Analysis in the Coed Nant Lolwyn study. Other details as Figure 9.32.

Table 9.17 Coed Nant Lolwyn, normal Indicator Species Analysis. The percentage occurrence of each species within each association.

	Association					
	A	B	C	D	E	F
Agrostis capillaris	0	0	0	0	14	12
Anthoxanthum odoratum	0	0	1	0	21	0
Arrhenatherum elatius	0	0	6	11	0	0
Brachypodium sylvaticum	0	2	9	11	28	0
Dactylis glomerata	0	0	6	0	42	0
Festuca gigantea	0	0	0	0	21	0
Holcus mollis	3	4	10	22	78	100
Luzula campestris	0	0	0	0	14	25
Melica uniflora	3	11	22	0	14	12
Milium effusum	0	4	19	0	0	0
Poa nemoralis	5	2	12	0	0	0
Poa trivialis	1	9	10	0	14	12
Adoxa moschatellina	1	0	30	11	7	0
Anemone nemorosa	81	88	74	22	100	75
Chrysosplenium oppositifolium	10	4	6	0	0	0
Circaea lutetiana	13	68	42	0	28	0
Conopodium majus	0	0	4	11	57	12
Epilobium montanum	0	2	9	0	0	0
Filipendula ulmaria	8	13	3	0	14	0
Fragaria vesca	0	6	3	0	7	0
Galium aparine	61	54	87	88	7	12
Geranium robertianum	8	31	33	33	7	0
Geum urbanum	18	70	36	11	21	0
Glechoma hederacea	0	4	4	55	0	0
Heracleum sphondylium	5	3	7	0	21	0
Hyacinthoides non-scripta	33	50	48	11	35	62
Lapsana communis	0	2	4	0	0	0
Mercurialis perennis	86	45	10	22	7	0
Moehringia trinervia	0	0	4	44	7	0
Oxalis acetosella	8	20	60	11	57	75
Potentilla sterilis	0	2	21	0	7	0
Primula vulgaris	0	2	15	0	28	0
Ranunculus ficaria	61	88	80	11	85	0
Rumex crispus	0	0	3	11	0	0
Sanicula europaea	0	18	30	0	42	0
Silene dioica	1	4	31	66	14	0
Solidago virgaurea	0	0	0	0	28	0
Stachys betonica	0	0	0	0	14	25
Stachys sylvatica	1	2	3	11	0	0
Stellaria holostea	3	0	43	0	57	37

Table 9.17 (*continued*)

	A	B	C	D	E	F
			Association			
Taraxacum officinale	0	2	12	0	28	0
Urtica dioica	22	6	13	77	0	0
Veronica chamaedrys	0	11	22	0	57	0
Veronica hederifolia	10	4	3	0	0	0
Veronica montana	6	59	46	0	14	0
Viola riviniana	0	11	13	22	42	0
Acer pseudoplatanus	10	11	4	0	7	0
Crataegus monogyna	0	4	3	22	7	0
Fraxinus excelsior	1	34	7	0	21	12
Hedera helix	13	29	36	44	35	12
Lonicera periclymenum	0	4	4	0	0	12
Prunus spinosa	1	6	4	0	0	0
Rosa canina	3	4	4	0	7	0
Rubus fruticosus	10	2	30	11	21	0
Rubus idaeus	0	0	4	0	0	0
Atrichum undulatum	6	6	4	0	7	0
Brachythecium rutabulum	6	0	9	44	7	0
Brachythecium velutinum	3	0	5	0	0	0
Cirriphyllum piliferum	1	2	1	0	21	0
Eurhynchium praelongum	61	61	77	100	57	25
Eurhynchium striatum	49	47	42	22	35	62
Fissidens bryoides	1	0	3	11	0	0
Fissidens taxifolius	5	0	1	0	0	0
Lophocolea bidentata	6	2	10	0	0	12
Mnium hornum	6	0	16	0	7	37
Plagiochila asplenioides	13	20	9	0	21	62
Plagiomnium undulatum	30	29	25	0	21	12
Plagiothecium denticulatum	0	2	6	0	0	12
Plagiothecium nemorale	10	4	6	22	0	12
Polytrichum formosum	1	0	0	0	0	37
Rhynchostegium confertum	5	11	6	0	0	0
Rhytidiadelphus loreus	0	0	1	0	7	25
Thamnobryum alopecurum	3	4	1	0	0	0
Thuidium tamariscinum	25	11	9	0	21	75
Athyrium filix-femina	3	2	0	0	0	0
Dryopteris dilatata	10	11	9	0	0	0
Dryopteris filix-mas	0	4	6	11	0	0

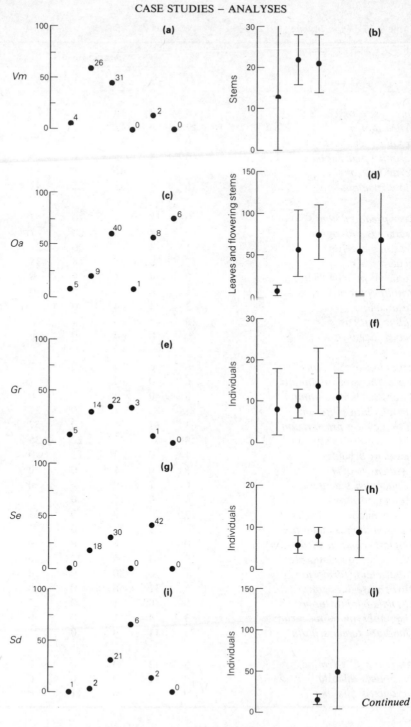

Continued

304

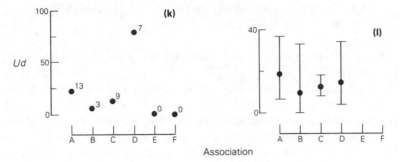

Figure 9.34 Percentage frequencies (a, c, e, g, i, k) and densities, m^{-2}, (b, d, f, h, j, l) of: (a, b) *Veronica montana*, (c, d) *Oxalis acetosella*, (e, f) *Geranium robertianum*, (g, h) *Sanicula europaea*, (i, j) *Silene dioica*, (k, l) *Urtica dioica*; in associations produced by Indicator Species Analysis in the Coed Nant Lolwyn study. Other details as Figure 9.32.

Comparison of the monothetic and polythetic classifications

From many points of view, the two classificatory methods have produced similar results. With the usual stopping rule of $\sqrt{n}$, where n is the number of stands in the whole data set, the normal Information Statistic classification produced six associations. The normal Indicator Species Analysis was therefore also adjusted to give six associations, but there was more than one way of achieving this. One obvious course of action was to not subdivide the 22 stands on the positive side of the first dichotomy any further; but in the analysis already presented we have seen that there are important environmental and vegetational differences between the two associations, E and F, formed from these 22 stands. Allowing the 66 stands of Association C to subdivide gave the very unequal split of 65 and 1. Although the one stand, no. 186, was rather unique in its floristic composition, it was felt that this division hardly contributed anything to the overall picture. Finally, subdivision of Association A and/or B did not produce new associations with notably different environmental or vegetational features. Accordingly, the six-association normal Indicator Species Analysis, presented above, was decided upon.

Given the two sets of results as presented, the following criteria can be applied to assess the efficacy of the two classifications:

(a) The magnitude of the differences between the mean levels of the environmental factors of the associations in relation to an approximate least significant range.
(b) The magnitude of differences of species compositions (and abundances) between associations.

305

(c) The distribution of associations on the ground in relation to the general field environment.

For Coed Nant Lolwyn, Indicator Species Analysis appears to give slightly superior results in respect of the first criterion (Table 9.18). However, it is not possible to distinguish between the two classifications on the basis of the second criterion. Both sets of associations define the kinds of species assemblages one might expect to find on the floor of a deciduous wood. The third criterion is attempting to examine the stand groupings from the viewpoint of the whole environment, rather than just from the environmental factors actually measured in each stand. Thus this criterion is rather subjective and intuitive, and the reasons for preferring one classification to the other are not easy to state. By mapping the occurrences of the associations on Figure 4.7, it appeared to me that the associations produced by Indicator Species Analysis were slightly better than those of the Information Statistic classification.

Table 9.19 attempts to indicate the similarities between the two classifications. The third column of the table shows the number of stands occurring in a particular Indicator Species Analysis association (shown in column 1) that occur in each of the Information Statistic Analysis associations (shown in column 2). The sum of all the numbers in one block of

Table 9.18 The residual mean squares and F-ratios of the ANOVAs of the environmental factors based on associations produced by Information Statistic and Indicator Species Analysis (on qualitative data) of the Coed Nant Lolwyn data. Asterisk code as Table 9.10.

	Residual mean squares		F-ratio	
	Information Statistic	Indicator Species	Information Statistic	Indicator Species
soil pH	0.1497	0.1529	14.28***	13.17***
soil phosphorus	0.1469	0.1506	2.78*	1.78
soil potassium	0.3548	0.3609	3.18**	2.46*
soil magnesium	0.1176	0.1138	3.36*	4.76***
soil manganese	0.3490	0.3297	4.20**	6.70***
soil calcium	0.2849	0.2341	8.23***	18.43***
soil sodium	0.4871	0.4440	2.21	6.18***
soil organic matter	0.6811	0.6904	1.72	1.17
soil moisture	37.878	35.106	0.86	3.99**
pre-vernal light	0.2514	0.2650	2.70*	0.56
vernal light	0.2321	0.2460	3.77**	1.36
aestival light	0.4424	0.4671	4.14**	1.86

Table 9.19 Comparison of stand occurrences in the associations produced by normal Indicator Species Analysis and normal Information Statistic.

Indicator Species association (1)	Information Statistic association (2)	Number of stands of (1) in (2)	Percentage of number of stands of (1) in (2)	Percentage of number of stands of (2) in (1)
A	A	33	56	59
	B	18	31	72
	C	0	0	0
	D	3	5	4
	E	3	5	21
	F	2	3	10
B	A	16	36	29
	B	4	9	16
	C	2	5	14
	D	21	48	30
	E	1	2	7
	F	0	0	0
C	A	5	8	9
	B	2	3	8
	C	4	6	29
	D	44	67	62
	E	4	6	29
	F	7	10	35
D	A	1	11	2
	B	1	11	4
	C	0	0	0
	D	0	0	0
	E	1	11	7
	F	6	67	30
E	A	1	7	2
	B	0	0	0
	C	8	57	57
	D	3	21	4
	E	0	0	0
	F	2	14	10
F	A	0	0	0
	B	0	0	0
	C	0	0	0
	D	0	0	0
	E	5	63	36
	F	3	37	15

column 3 equals the total number of stands in the relevant Indicator Species Analysis association (column 1). Thus for Association A of Indicator Species Analysis, the relevant sum in column 3 is $33 + 18 + 0 + 3 + 3 + 2 = 59$, which is the total number of stands in that association. Column 4 expresses the numbers in column 3 as a percentage of the total number of stands in the relevant Indicator Species Association. For example, in the first row of the table, 56% of the stands in Indicator Species Association A occur in Information Statistic Association A. The last column of the table is the converse; again in the first row of the table, 59% of the stands in Information Statistic Association A occur in Indicator Species Association A.

Looking at the table as a whole, one may conclude that all the stands of Association F of Indicator Species Analysis are divided between Associations E and F of the Information Statistic classification, with a preponderance in Association E. More than half the stands of Indicator Species Association E occur in Information Statistic Association C and vice-versa. Two thirds of the stands of Indicator Species Association D occur in Information Statistic Association F, and a similar situation exists between Indicator Species Association C and Information Statistic Association D. The stands of Indicator Species Association A mostly occur in Information Statistic Associations A and B, while stands of Indicator Species Association B are mostly divided between Information Statistic Associations D and A.

Thus we may identify similar associations between the two classification techniques – similar, but by no means identical – and these related associations are summarized in Table 9.20.

Table 9.20 'Corresponding' associations produced by Indicator Species Analysis and Information Statistic Analysis, based on information contained in Table 9.19.

Indicator Species association	A	B	C	D	E	F
Information Statistic association	A(B)	D(A)	D	F	C	E

Inverse classifications

Inverse Information Statistic and inverse Indicator Species classifications are shown in Figures 9.35 and 9.36, respectively. The former, with the usual stopping rule (square root of the total number of species), yields 18 associations. By allowing not more than five levels of division, the inverse Indicator Species Analysis was made to yield 17 associations. However, the composition of the associations in each of the two classification schemes is almost entirely different (Tables 9.21 & 9.22). Only one grouping is

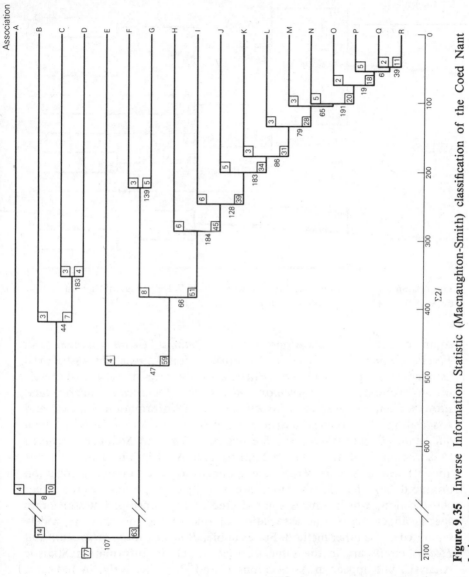

Figure 9.35 Inverse Information Statistic (Macnaughton-Smith) classification of the Coed Nant Lolwyn study.

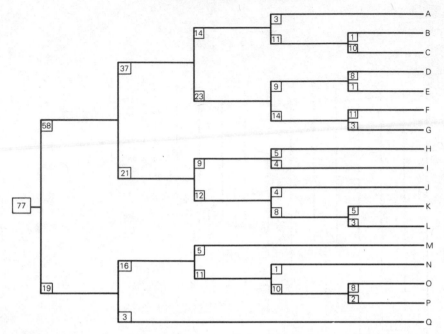

Figure 9.36 Inverse Indicator Species Analysis (TWINSPAN) of the qualitative data of the Coed Nant Lolwyn study.

common to both classifications: *Circaea lutetiana*, *Geum urbanum*, and *Veronica montana* (Association C in Information Statistic and Association G in Indicator Species). Otherwise there are only traces of associated species within groupings, e.g. *Anemone nemorosa*, *Ranunculus ficaria*, and *Eurhynchium praelongum* (Association A of Information Statistic and Association C of Indicator Species); *Agrostis capillaris*, *Anthoxanthum odoratum*, *Luzula campestris*, *Stachys betonica*, and *Solidago virgaurea* (Association P of Information Statistic and Association O of Indicator Species). Some species which are grouped together in one association produced by one classification, are in adjacent groups in the other classification, which implies some degree of relationship; but sometimes species appearing in one association of one of the methods are widely separated in the other method. For example, *Brachypodium sylvaticum* and *Holcus mollis* are in the same association, G, in Information Statistic Analysis, but appear in Associations K and P, respectively, in Indicator Species Analysis. These two associations are on opposite sides of the primary division!

Faced with these very divergent results produced by the two different inverse classifications, one is left wondering what practical uses and what

Table 9.21 List of species in each association of the inverse Information Statistic of the Coed Nant Lolwyn data.

Association	Number of species	List of species
A	4	An, Ga, Rf, Ep
B	3	Hn, Mp, Es
C	3	Cl, Gu, Vm
D	4	Mu, Pn, Vc, Pu
E	4	Oa, Hh, Pa, Tt
F	3	Gr, Se, Sh
G	5	Bs, Hm, Cm, Pv, Fe
H	6	Me, Am, Sd, Ud, Ru, Cp
I	6	Pt, Co, Fu, Hs, Vh, Rt
J	5	Vr, Ap, Au, Lb, Pl
K	3	Ps, Mh, Dd
L	3	Gh, Mt, Br
M	3	Fv, To, Cg
N	5	Dg, Em, Lp, Pp, Rc
O	2	Bv, Df
P	5	Ac, Ao, Lz, Sb, Sv
Q	2	Ae, Fb
R	11	Fg, Lc, Rx, Ss, Ri, Ft, Pd, Pf, Rl, Ta, Af

Table 9.22 List of species in each association of the inverse Indicator Species Analysis of the Coed Nant Lolwyn (qualitative) data.

Association	Number of species	List of species
A	3	Fu, Fv, Fe
B	1	Au
C	10	An, Hn, Rf, Hh, Rc, Ep, Es, Lb, Pu, Pn
D	8	Co, Mp, Vh, Pp, Ft, Ta, Af, Dd
E	1	Ud
F	11	Pn, Ga, Gr, Gh, Lc, Ss, Ps, Bv, Rt, Fb, Df
G	3	Cl, Gu, Vm
H	5	Ae, Me, Em, Rx, Ri
I	4	Am, Mt, Ps, Sd
J	4	Mu, Ru, Cg, Br
K	5	Bs, Pv, Sh, To, Pd
L	3	Oa, Se, Mh
M	5	Pt, Hs, Vc, Vr, Lp
N	1	Dg
O	8	Ac, Ao, Fg, Lz, Sb, Cm, Sv, Cp
P	2	Hm, Rl
Q	3	Pa, Pf, Tt

ecological justifications there are for such classifications in relation to the Coed Nant Lolwyn data. The answer must be, very few. Results such as these tend to emphasize the continuum nature of the ground flora of Coed Nant Lolwyn, and any attempts to obtain species groupings is quite artificial. This study also demonstrates the value of applying more than one classification method to a data set. If certain species have a high propensity to occur together and so can be regarded as a real ecological group, then those species should be grouped together regardless of the classification method. In Coed Nant Lolwyn, only *Circaea lutetiana*, *Geum urbanum* and *Veronica montana* show this propensity. In the Iping Common transect, many more species groupings were common to the two classification methods, but there were only a few groups involved which obviously assists in this respect.

Detrended Correspondence Ordination

The first two axes of the stand and species Detrended Correspondence Ordinations are shown in Figures 9.37 and 9.38, respectively, and should be examined in conjunction with Table 9.23 which gives Spearman's rank correlation coefficients of levels of the measured environmental factors with ordination axis scores. From Table 9.23 it is clear that the first axis essentially represents a base status gradient – high on the left, low on the

Table 9.23 Spearman's rank correlation coefficients of qualitative Detrended Correspondence Ordination axis scores and environmental factors for the Coed Nant Lolwyn data. Asterisk code of significance levels as Table 9.10.

	Axis			
	1	2	3	4
soil pH	-0.26^{***}	-0.13	0.05	0.02
soil phosphorus	0.12	0.26^{***}	0.09	0.27^{***}
soil potassium	-0.06	0.01	0.19^{**}	0.09
soil magnesium	-0.17^{*}	-0.04	0.14^{*}	0.02
soil manganese	0.20^{**}	0.18^{*}	0.26^{***}	0.30^{***}
soil calcium	-0.26^{***}	-0.09	0.08	0.04
soil sodium	-0.19^{**}	-0.06	0.15^{*}	0.07
soil organic matter	0.01	0.05	0.14	0.11
soil moisture	-0.04	0.02	0.20^{**}	0.13
pre-vernal light	-0.17^{*}	-0.05	0.05	0.01
vernal light	-0.00	-0.03	0.15^{*}	0.06
aestival light	-0.07	0.03	0.18^{*}	0.10

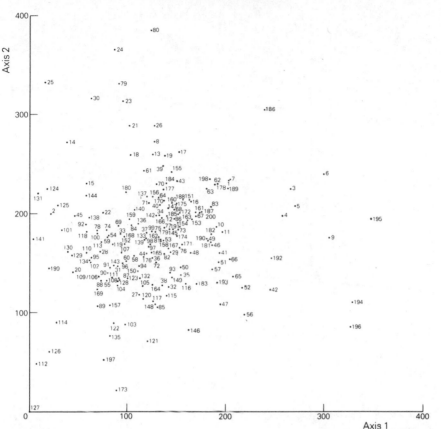

Figure 9.37 Detrended Correspondence stand ordination of the Coed Nant Lolwyn study.

right – and the second axis represents a decreasing soil phosphorus content gradient from top to bottom. None of the correlation coefficients is particularly high, but the largest (in the absolute sense) are all highly significantly different from zero ($P \leqslant 0.001$). Axes 3 and 4 do not appear to provide any significantly extra information: Axis 4 seems to be a repetition of Axis 2, while soil manganese is highly correlated with Axis 3 scores.

The species positions on the ordination diagram (Fig. 9.38) reinforce conclusions drawn on the basis of the classification results, discussed earlier. Species such as *Mercurialis perennis*, *Urtica dioica*, *Geum urbanum*, and *Galium aparine* are at the base-rich end of Axis 1, while *Luzula campestris*, *Holcus mollis*, *Oxalis acetosella*, and *Stellaria holostea* lie towards the base-poor end. Similarly on Axis 2, *Moehringia trinervia*, *Glechoma hederacea*, *Silene dioica*, and *Urtica dioica* are at the upper,

313

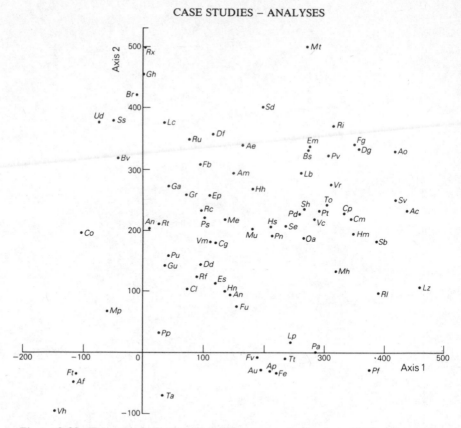

Figure 9.38 Detrended Correspondence species ordination of the Coed Nant Lolwyn study.

phosphorus-rich, end of the axis; while many of the bryophytes and *Mercurialis perennis* are at the oppostie end.

The position of *Veronica hederifolia* is peculiar, occurring at the extreme ends of both the first and second axes (also of Axis 3), implying that this species grows in base-rich but phosphorus-poor soils in Coed Nant Lolwyn. However, further analysis of the occurrence of *Veronica hederifolia* in relation to these soil factors fails to substantiate these suggestions. Remember that not only are the correlation coefficients involved rather low, but that *Veronica hederifolia* occurs in only 10 stands out of the 200 in the data set. Clearly, trends highlighted by the ordination results must also be examined in other ways.

Other ordination results

There is little point in attempting a Bray & Curtis Ordination here; the number of stands is large, and great subjectivity from prior knowledge would be required to designate the end points of the axes. However, Reciprocal Averaging and the variants of Principal Component Ordination do yield interesting results. Tables 9.24–9.26 show details for each of the first three axes (second to fourth in the case of the non-centred PCOs), respectively, of the species with the highest loadings and the most significant environmental correlations for a variety of ordination methods.

The outputs from the programs for RA and DCO do not give species loadings in the form described for PCO, but species scores have been converted to loadings in the following way. First, the scores were centred about the origin by finding their mean and then subtracting it from each score. Secondly, the sum of squares of the new scores was obtained. Finally each new species score was squared, divided by the sum of squares of the new scores and the square root of the result obtained; the result is the species loading required, and the method of calculation ensures that the sum of squares of the species loadings sum to unity, as in the case of the PCO loadings.

Let us now examine closely the section for DCO in Table 9.24. Soil pH and calcium have a highly significant negative correlation ($P < 0.001$) with Axis 1 stand scores, while sodium and manganese have significant correlations ($P < 0.01$), with manganese being positively correlated. This means that soil pH, calcium and sodium levels tend to *decrease* as one moves along Axis 1 from left to right (increasing scores) while soil extractable manganese levels tend to *increase* with increasing Axis 1 scores. *Luzula campestris* and *Agrostis capillaris*, being highly positively loaded onto Axis 1, implies that they are species associated with high soil manganese levels and/or low soil sodium, calcium, pH. *Veronica hederifolia*, *Athyrium filix-femina*, *Fissidens taxifolius*, and *Chrysosplenium oppositifolium* are implicated as having the opposite properties, that is, associated with soils of high pH, calcium, sodium and/or low manganese levels.

It is, however, very important to remember that statements such as those just made are no more than *possibilities*; they are not proven facts. The correlations, although very significantly different from zero, are nowhere near perfect (± 1); also the maximum loadings are fairly small. Any ideas generated by this kind of approach must be checked by other, more direct, methods. For the present data we use the contingency table method of examining the species frequencies in stands whose mean levels of an environmental factor is on the one hand greater than, and on the other hand less than, the mean level of that factor in all the stands (demonstrated in Example 8.4). Further, graphs like that shown in Figure 8.2 were drawn to show the distribution of the occurrence of each species against each environmental factor. Such graphs were examined qualitatively to give extra information.

Chi-square values from the contingency tables are given in the first part

315

Table 9.24 Loadings of species on, and correlation coefficients of environmental factors with, the first axis (second axis in the case of the non-centred versions of Principal Component Ordination) of a variety of ordination methods. Only significant ($P < 0.01$) environmental correlations and species whose loadings exceed 0.2 in absolute value are shown. Loadings for Reciprocal Averaging and Detrended Correspondence Ordinations are calculated from the conventional species scores by the method described in the text. Asterisk code of significance levels as Table 9.10. .... separates positively and negatively loaded species.

Principal Component Ordinations

non-centred, non-standardized				non-centred, standardized			
species		environmental factors		species		environmental factors	
Mp	0.57	Ca	0.38***	....		Mg	0.31***
Ep	0.20	pH	0.34***	Hm	−0.20	Ca	0.30***
....		Mg	0.33***	Rl	−0.22	K	0.30***
Se	−0.26	Na	0.32***	Cm	−0.24	H₂O	0.29***
Sh	−0.29	L3	0.30***	Sb	−0.28	L1	0.28***
Oa	−0.33	K	0.28***	Dg	−0.28	Na	0.27***
		H₂O	0.28***	Lz	−0.29	L3	0.25***
		L1	0.28***	Sv	−0.32	pH	0.25***
		Org	0.27***	Ao	−0.33	Org	0.24***
		P	0.25***	Ac	−0.35		
		L2	0.18**				

centred, non-standardized				centred, standardized			
species		environmental factors		species		environmental factors	
Cl	0.38	Mn	0.18**	Dg	0.29	Mn	0.19**
Gu	0.35			Ac	0.28		
Vm	0.34			Sv	0.27		
Rf	0.33			Ao	0.26		
Se	0.25			Cm	0.26		
Sh	0.25			Sb	0.25		
Gr	0.23			Hm	0.22		
Oa	0.21			Lz	0.21		
....				Sh	0.20		
Mp	−0.28			Vr	0.20		
				....		Ca	−0.24***
				Mp	−0.22	pH	−0.26***

Table 9.24 (*continued*)

Other methods

Reciprocal Averaging Ordination		Detrended Correspondence Ordination	
species	environmental factors	species	environmental factors
Lz 0.47	Mn 0.19**	Lz 0.22	Mn 0.20**
Ac 0.40		Ac 0.21	
Ao 0.34		....	
Sv 0.34		Co −0.21	
Rl 0.24	Na −0.21**	Ft −0.22	Na −0.19**
Sb 0.23	Ca −0.27***	Af −0.22	Ca −0.26***
....	pH −0.28***	Vh −0.25	pH −0.26***

of Table 9.27 for the relevant species and environmental factors highlighted by the first axis of DCO (Table 9.24). This method of ordination seems to have 'pulled out' the rare species in this data set, which limits the magnitude of χ^2 regardless of the degree of association. For example, the χ^2-values for soil calcium and both the species *Luzula campestris* and *Agrostis capillaris* are the maximum possible: there are no occurrences of either of these two species in stands whose soil calcium levels are above the general mean level of this element in all stands.

The common feature of both *Luzula campestris* and *Agrostis capillaris* is their significant negative association with soil calcium. In other respects they differ: *Luzula campestris* is almost significantly associated with soils of low pH, but *Agrostis capillaris* appears indifferent. However, further examination reveals that the three stands in which the latter species occurs have soil pH-values of 4.0, 4.7 and 4.9; and the mean pH for all stands is 4.85. Clearly *Agrostis capillaris* tends towards soils of lower pH in this data set.

Of the remaining species, except for *Veronica hederifolia*, low soil manganese and high soil pH levels seem to be the important factors, with high levels of soil calcium and sodium being less important. The position of *Veronica hederifolia* is quite anomalous; it must be due to some artefact of the ordination method. The position of this species at the extreme end of Axis 1 is completely misleading as it has no significant associations with any of the soil factors currently under discussion. Its distribution scans large ranges of soil pH and manganese levels, a mid-range of sodium levels, but it does not occur in any stand having a calcium value of less than 60 mg $(100 \text{ g})^{-1}$. Indeed, *Veronica hederifolia* does not have significant associations with any of the measured environmental factors, but there is a tendency towards stands of higher pre-vernal and vernal light intensities, as

317

might be expected of a winter annual that flowers and sets seed in the spring. However, this example does show the care that must be taken in the interpretation of species loadings on, and significantly correlated environmental factors with, ordination axes.

Turning now to the Reciprocal Averaging Axis 1, we see that the environmental factor correlations are virtually identical with those in DCO, but there are some differences among the most highly loaded species. There are now no species with a loading of less than -0.2, but there are several more species with loadings greater than $+0.2$. We may also examine the

Table 9.25 Loadings of species on, and correlation coefficients of environmental factors with, the second axis (third axis in the case of the non-centred versions of Principal Component Ordination) of a variety of ordination methods. Other details as Table 9.24.

Principal Component Ordinations

non-centred, non-standardized				non-centred, standardized		
species		environmental factors		species		environmental factors
Es	0.33	P	0.29***	Bs	0.24	Mn 0.26***
Oa	0.32			Se	0.21	
En	0.26			. . . .		
Hm	0.25			Tt	-0.24	
Tt	0.23			Pa	-0.27	
Pa	0.22			Pf	-0.33	
. . . .						
Cl	-0.32					
Rf	-0.33					
Gu	-0.35					

centred, non-standardized				centred, standardized		
species		environmental factors		species		environmental factors
Mp	0.41	Ca	0.44***	Rf	0.28	Mn 0.19**
Gu	0.28	Mg	0.41***	Cl	0.26	
Rf	0.25	Na	0.40***	Gu	0.25	
Cl	0.23	pH	0.38***	Vm	0.24	
Ga	0.23	K	0.37***	Se	0.23	
. . . .		L3	0.36***	Ps	0.22	
Hm	-0.29	H$_2$O	0.35***	Gr	0.21	
Oa	-0.43	L1	0.32***	Sh '	0.20	
		Org	0.30***	. . . .		
		L2	0.23**	Lz	-0.23	

Table 9.25   (*continued*)

Other methods

Reciprocal Averaging Ordination		Detrended Correspondence Ordination	
species	environmental factors	species	environmental factors
Pf 0.34	Ca 0.32***	Mt 0.24	P 0.26***
Lz 0.28	H$_2$O 0.32***	Rx 0.24	
Ta 0.23	Org 0.31***	Gh 0.21	
Vh 0.23	K 0.31***	....	
Af 0.20	Na 0.30***	Ft − 0.20	
....	L3 0.30***	Af − 0.21	
Gh − 0.21	Mg 0.30***	Ta − 0.22	
Mt − 0.26	pH 0.28***	Vh − 0.25	
Rx − 0.32	L1 0.24***		
	L2 0.23**		
	Mn 0.20**		

most reliable (theoretically speaking) of the PCOs – the centred and standardized version. Apart from the disappearance of soil sodium as a factor significantly correlated with Axis 1, the environmental interpretation of this axis is similar to the foregoing. However, many more species are now loaded on Axis 1 to an absolute value greater than 0.2, including some having much higher frequencies in the data set than those already considered. The new species results are shown in the second part of Table 9.27, and you may readily draw your own conclusions.

The results for the first axis of the centred and non-standardized PCO make very little sense. Soil manganese is the only significant environmental correlation, and that not very strong. *Mercurialis perennis* is indeed strongly negatively associated with this factor; but of the species supposedly positively associated with soil manganese, only *Oxalis acetosella* is significantly so ($\chi^2 = 6.35^*$), while *Geum urbanum, Sanicula europaea, Stellaria holostea* and *Geranium robertianum* are only very weakly positively associated with soil manganese ($\chi^2 = 1.53, 0.88, 0.68, 0.08$, respectively). *Ranunculus ficaria* is slightly *negatively* associated with soil manganese ($\chi^2 = 0.66$), while *Veronica montana* is *significantly* negatively associated with this factor ($\chi^2 = 3.85^*$)!

The results produced by the non-centred PCOs are not very helpful in that so many environmental factors are significantly correlated with Axis 2. In the case of the non-standardized analysis, we can infer that stands with high positive scores on this axis tend to have base-rich soils. *Mercurialis perennis* has a very high positive loading on Axis 2, and confirms what we

319

Table 9.26 Loadings of species on, and correlation coefficients of environmental factors with, the third axis (fourth axis in the case of the non-centred versions of Principal Component Ordination) of a variety of ordination methods. Other details as Table 9.24.

Principal Component Ordinations

non-centred, non-standardized			non-centred, standardized		
species		environmental factors	species		environmental factors
Ga	0.40	P 0.19**	Gh	0.34	P 0.34***
Ep	0.34		Mt	0.32	Mn 0.18**
Sd	0.31		Sd	0.28	
Ud	0.24		Ud	0.25	
....			Ru	0.22	
Mp	−0.20		....		
En	−0.27		Fe	−0.21	
An	−0.36		Ps	−0.21	
			Fv	−0.25	

centred, non-standardized			centred, standardized		
species		environmental factors	species		environmental factors
Es	0.59	P 0.23***	Fe	0.24	H_2O 0.25***
Ep	0.38	K 0.22**	An	0.24	K 0.22**
Pu	0.32	Mg 0.22**	....		L3 0.20**
Vm	0.29	Org 0.20**	Ep	−0.21	Org 0.20**
....		H_2O 0.19**	Ga	−0.21	Mn 0.20**
			Ru	−0.21	Na 0.19**
			Ud	−0.23	Mg 0.18**
			Gh	−0.24	Ca 0.18**
			Mt	−0.25	L2 0.18**
			Sd	−0.33	

Other methods

Reciprocal Averaging Ordination			Detrended Correspondence Ordination		
species		environmental factors	species		environmental factors
Mt	0.37	P 0.27***	Vh	0.20	Mn 0.26***
Lz	0.28		....		K 0.19**
Br	0.26		Pn	−0.23	
Gh	0.24		Dd	−0.24	
....			Pd	−0.26	
Fv	−0.22		Ta	−0.27	
Vh	−0.25				

Table 9.27 Chi-square values obtained from contingency tables based on species frequencies in stands having levels of an environmental factor greater than or less than the mean level for all stands. The sign in brackets indicates whether the species has a higher frequency in stands with high levels of the environmental factor than expected on the basis of random occurrence (positive association, $+$), or vice versa (negative association, $-$). Asterisk code of significant levels as in Table 9.10.

	Frequency	x^2 Mn	Na	Ca	pH
Luzula campestris	4	1.15($-$)	0.15($-$)	5.87*($-$)	3.77($-$)
Agrostis capillaris	3	0.28($+$)	0.87($-$)	4.38*($-$)	0.26($-$)
Chrysosplenium opp.	13	10.68**($-$)	1.75($+$)	1.85($+$)	4.66*($+$)
Fissidens taxifolius	4	4.33*($-$)	2.78($+$)	2.84($+$)	4.42*($+$)
Athyrium filix-fem.	3	3.23($-$)	2.07($+$)	2.12($+$)	3.30($+$)
Veronica hederifol.	10	0.30($+$)	0.00($+$)	0.35($-$)	1.37($-$)
Dactylis glomerata	10	6.25*($+$)	1.66($-$)	10.45**($-$)	0.61($+$)
Solidago virgaurea	4	0.90($+$)	0.15($-$)	5.87*($-$)	0.01($+$)
Anthoxanthum odor.	4	0.90($+$)	0.15($-$)	5.87*($-$)	0.87($-$)
Rhytidiadelphus lor.	4	0.00($-$)	6.00*($-$)	1.95($-$)	0.87($-$)
Stachys betonica	4	0.90($+$)	2.02($-$)	5.87*($-$)	0.87($-$)
Conopodium majus	13	6.10*($+$)	2.55($-$)	10.93***($-$)	3.46($-$)
Holcus mollis	32	3.04($+$)	2.52($-$)	12.13***($-$)	2.83($-$)
Stellaria holostea	42	0.68($+$)	1.99($-$)	7.54**($-$)	3.22($-$)
Viola riviniana	22	2.75($+$)	5.11*($+$)	3.34($-$)	0.06($-$)
Mercurialis peren.	81	15.63***($-$)	2.90($+$)	19.84***($+$)	30.39***($+$)

already know about this species. What about *Eurhynchium praelongum*? Does it also tend to occur on base-rich soils? A similar investigation to the above shows that only soil magnesium is significantly positively associated with the occurrence of *Eurhynchium praelongum* ($x^2 = 6.09^*$); however, vernal light (L2) is highly significantly negatively associated with this species ($x^2 = 8.76^{**}$) even though L2 is significantly positively correlated with Axis 2! Among the negatively loaded species on Axis 2, *Oxalis acetosella* is certainly usefully predicted as occurring on acid, base-poor soils (x^2-values: $32.41^{***}(-)$ for pH, $25.54^{***}(-)$ for Ca, $13.54^{***}(-)$ for Mg, and $11.22^{***}(-)$ for Na). *Stellaria holostea*, however, is only significantly negatively associated with soil calcium ($x^2 = 7.54^{**}$), and *Sanicula europaea* is negatively associated with soil phosphorus ($x^2 = 4.53^*$).

There are no new species with high loadings in the standardized non-centred PCO. Soil magnesium is negatively associated with all the species, as predicted, but only the result with *Rhytidiadelphus loreus* is significant ($x^2 = 5.09^*$). On the other hand, *Holcus mollis* has a fairly high *positive* association with soil potassium ($x^2 = 3.72$) whereas the PCO result indicates that the association should be negative.

321

Turning now to the second (third) axis (Table 9.25), we will confine attention to those ordinations giving just a single significantly correlated environmental factor. In the case of DCO, where soil phosphorus is significantly correlated with Axis 2 scores, the most positively loaded species are only weakly positively associated with soils of above average phosphorus status (χ^2-values: 0.91 for *Moehringia trinervia*, 0.52 for *Rumex crispus*, 0.83 for *Glechoma hederacea*). However, a glance at Figure 9.38 shows other species with only slightly lower loadings (or scores) on Axis 2 that do, in fact, show much stronger affinities with soils of high phosphorus status, e.g. *Silene dioica* ($\chi^2 = 5.91^*$) and *Luzula campestris* ($\chi^2 = 4.79^*$), but *Urtica dioica* is much weaker in this respect ($\chi^2 = 0.25$). The most negatively loaded species on Axis 2 are also not clear cut in their association with soils of low phosphorus status; whereas the species which actually do show large negative associations with such soils – *Anemone nemorosa* ($\chi^2 = 11.37^{***}$), *Ranunculus ficaria* ($\chi^2 = 7.60^{**}$), *Circaea lutetiana* ($\chi^2 = 4.59^*$) and *Sanicula europaea* ($\chi^2 = 4.53^*$) – do not occur conspicuously towards the negative end of Axis 2 (Fig 9.38). Despite the significant correlation of soil phosphorus status with Axis 2, this axis actually represents more of a wood edge to interior gradient, as already discussed (p. 234). If each stand's distance from the southern woodland boundary was correlated with Axis 2 scores, a higher coefficient than 0.26 might result.

The second axis of the centred and standardized PCO presents a very similar picture to that of the first axis of the centred and non-standardized PCO, with no useful species–environment relationships suggested. The third axis of the non-centred and non-standardized PCO, showing soil phosphorus with a high correlation, has species with a high negative loading which actually are negatively associated with low phosphorus: *Geum urbanum* ($\chi^2 = 3.40$), together with *Ranunculus ficaria* and *Circaea lutetiana*; but *Holcus mollis* is the only positively loaded species with a positive association with soil phosphorus ($\chi^2 = 5.91^*$). In the case of the third axis of the non-centred and standardized PCO, only *Brachypodium sylvaticum* (wood false-brome) ($\chi^2 = 3.10$, positive association with soil manganese) has the correct inferred relationship, but it is not statistically significant.

Finally, we consider briefly the third (fourth) axis (Table 9.26). That for DCO turns out to be uninformative, but for RA ordination *Luzula campestris* does indeed have a significant positive association with soil phosphorus ($\chi^2 = 4.79^*$). The centred PCOs have too many environmental factors significantly correlated with Axis 3 for easy interpretation, so we are left with the fourth axes of the non-centred PCOs. In the non-standardized version only *Anemone nemorosa* is significantly associated with soils of low phosphorus status. In the standardized analysis, *Silene dioica* is positively associated with soils of high phosphorus, and *Moehringia trinervia* with high managanese ($\chi^2 = 4.32^*$).

Concluding remarks on vegetation analysis results

Vegetation is a complex collection of plant species. The interaction of plants of different species with each other, and with their environment, ensures that attempts to unravel species–environment relationships are bound to be difficult. Rarely will there be clear-cut relationships, more usually the picture will be indistinct.

The methods of vegetation analysis can only be regarded as providing possible initial pointers to important species–environment relationships. The results produced by analytical methods are only as good as the data collected in the field; sophisticated analyses cannot refine poor data, hence the importance of good field work, including the devising of the sampling scheme, good species identification and accuracy in any subsequent laboratory work.

Even when the most efficient of modern analytical techniques are applied to good data, the results contain a mixture of real trends together with spurious ones generated by the underlying mathematical model which is never completely appropriate to the data in hand. The difficulty is in separating these two indicated trends, and this is the reason for the importance of undertaking further analytical work on the same data. For the Coed Nant Lolwyn study we took a rather simplistic approach by looking at only one environmental factor at a time. Where an ordination indicated the importance of several environmental factors simultaneously by their high correlations on an axis, better results might have been obtained in the follow-up if all the indicated environmental factors of importance had been examined together in some way.

Another feature of at least some vegetation analysis methods (classifications), examined in the Iping Common study but not reported here, is that their results seem somewhat 'unstable', in that small changes in the data may produce disproportionately greater changes in the results. The methods do not seem to be very robust. Again this is an indication that we must treat the results of vegetation analyses with some caution, as a first look at the vegetation itself, and of the relationships between the vegetation and its environment.

These two apparently undesirable propensities of vegetation analytical methods – the partial inapplicability of the underlying mathematical models and the tendency of a particular result being very closely allied to a particular set of data – should not, however, blind us to the usefulness of such methods. It is only by them that a reasonable summary of our field data can be achieved; only by using these methods can we find some sort of order in what is otherwise an indigestible mass of data. BUT, the results of a vegetation analysis are only the first step in this kind of scientific enquiry. Many ideas are generated by such results, but *they are only hypotheses*. Much more work, both in the field and laboratory, is needed to substantiate a few of these many hypotheses. Indeed, in our relatively brief human lifetimes it may only be possible for any one of us to investigate substantially a few of the myriad of possibilities suggested by a single well-executed habitat study.

References

Al-Mufti, M. M., C. L. Sydes, S. B. Furness, J. P. Grime, & S. R. Band, 1977. A quantitative analysis of shoot phenology and dominance in herbaceous vegetation. *Journal of Ecology* **65**, 759–91.

Allen, S. E., H. M. Grimshaw, J. A. Parkinson, & C. Quarmby, 1974. *Chemical analysis of ecological materials.* Oxford: Blackwell.

Anderson, M. C. 1964. Studies of the woodland light climate. II. Seasonal variation in the light climate. *Journal of Ecology* **52**, 643–63.

Ball, D. F. & W. M. Williams, 1968. Variability of soil chemical properties in two uncultivated brown earths. *Journal of Soil Science* **19**, 379–91.

Barker, H. & A. R. Clapham, 1939. Seasonal variations in the acidity of some woodland soils. *Journal of Ecology* **27**, 114–25.

Blackman, G. E. & A. J. Rutter, 1946. Physiological and ecological studies in the analysis of plant environment. I. The light factor and the distribution of the bluebell (*Scilla non-scripta*) in woodland communities. *Annals of Botany N.S.* **10**, 361–90.

Braun-Blanquet, J. 1927. *Pflanzensoziologie.* Wien: Springer.

Braun-Blanquet, J. 1932. *Plant sociology.* New York: McGraw-Hill.

Braun-Blanquet, J. 1951. *Pflanzensoziologie,* 2nd edn. Wien: Springer.

Bray, J. R. & J. T. Curtis, 1957. An ordination of the upland forest communities of southern Wisconsin. *Ecological Monographs* **27**, 325–49.

Brown, R. T. & J. T. Curtis, 1952. The upland conifer–hardwood forests of northern Wisconsin. *Ecological Monographs* **22**, 217–34.

Causton, D. R. 1983. *A biologist's basic mathematics.* London: Edward Arnold.

Causton, D. R. 1987. *A biologist's advanced mathematics.* London: Allen & Unwin.

Cottam, G. 1949. The phytosociology of an oak wood in south-western Wisconsin. *Ecology* **30**, 271–87.

Curtis, J. T. & R. P. McIntosh, 1950. The interrelation of certain analytic and synthetic phytosociological characters. *Ecology* **31**, 434–55.

Curtis, J. T. & R. P. McIntosh, 1951. An upland forest continuum in the prairie–forest border region of Wisconsin. *Ecology* **32**, 476–96.

Dagnelie, P. 1960. Contribution a l'étude des communautés végétales par l'analyse factorielle. *Bulletin du Service de la Carte Phytogéographique Sér.B* **5**, 7–71 & 93–195.

REFERENCES

Dahl, E. & E. Hadač, 1949. Homogeneity of plant communities. *Studia Botanica Cechoslovaca* 10, 159–76.

Davy, A. J. & K. Taylor, 1974. Seasonal patterns of nitrogen availability in contrasting soils in the Chiltern Hills. *Journal of Ecology* 62, 793–807.

Davy, A. J. & K. Taylor, 1975. Seasonal changes in the inorganic nutrient concentrations in *Deschampsia caespitosa* (L.) Beauv. in relation to its tolerance of contrasting soils in the Chiltern Hills. *Journal of Ecology* 63, 27–39.

Dony, J. G. 1967. *Flora of Hertfordshire*. Hitchin Urban District Council.

Duncan, D. B. 1955. Multiple range and multiple *F* tests. *Biometrics* 11, 1–42.

Edwards, A. W. F. & L. L. Cavalli-Sforza, 1965. A method for cluster analysis. *Biometrics* 21, 362–75.

Etherington, J. R. 1982. *Environment and plant ecology*, 2nd edn. Chichester: Wiley.

Fitter, A. H. & R. K. M. Hay, 1987. *Environmental physiology of plants*, 2nd edn. London: Academic Press.

Frankland, J. C., J. D. Ovington & C. Macrae, 1963. Spatial variations in soil, litter and ground vegetation in some Lake District woodlands. *Journal of Ecology* 51, 97–112.

Gauch, H. G. 1982. *Multivariate analysis in community ecology*. Cambridge: Cambridge University Press.

Gauch, H. G., R. H. Whittaker, & T. R. Wentworth, 1977. A comparative study of reciprocal averaging and other ordination techniques. *Journal of Ecology* 65, 157–74.

Goodall, D. W. 1952. Some considerations in the use of point quadrats for the analysis of vegetation. *Australian Journal of Scientific Research, Series B* 5, 1–41.

Goodall, D. W. 1953. Objective methods for the classification of vegetation. I. The use of positive interspecific correlation. *Australian Journal of Botany* 1, 39–63.

Goodall, D. W. 1954. Objective methods for the classification of vegetation. III. An essay in the use of factor analysis. *Australian Journal of Botany* 2, 304–24.

Greig-Smith, P. 1983. *Quantitative plant ecology*. Oxford: Blackwell.

Grime, J. P. 1979. *Plant strategies and vegetation processes*. Chichester: Wiley.

Harrison, A. F. 1975. Estimation of readily-available phosphate in some English Lake District woodland soils. *Oikos* 26, 170–6.

Harrison, C. M. 1970. The phytosociology of certain English heathland communities. *Journal of Ecology* 58, 573–89.

Hill, M. O. 1973. Reciprocal averaging: an eigenvector method of ordination. *Journal of Ecology* 61, 237–49.

Hill, M. O. 1974. Correspondence analysis: a neglected multivariate method. *Applied Statistics (Journal of the Royal Statistical Society, Series C)* 23, 340–54.

Hill, M. O. 1979a. *DECORANA – a FORTRAN program for detrended correspondence analysis and reciprocal averaging*. Ecology and Systematics, Cornell University, Ithaca, New York.

REFERENCES

Hill, M. O. 1979b. *TWINSPAN – a FORTRAN program for arranging multivariate data in an ordered two-way table by classification of the individuals and attributes*. Ecology and Systematics, Cornell University, Ithaca, New York.

Hill, M. O. & H. G. Gauch, 1980. Detrended correspondence analysis: an improved ordination technique. *Vegetatio* **42**, 47–58.

Hill, M. O., R. G. H. Bunce, & M. W. Shaw, 1975. Indicator species analysis, a divisive polythetic method of classification, and its application to a survey of native pinewoods in Scotland. *Journal of Ecology* **63**, 597–613.

Hubbard, C. E. 1984. *Grasses*, 3rd edn. London: Penguin.

Huntley, B. & H. J. B. Birks, 1979. The past and present vegetation of the Morrone Birkwoods National Nature Reserve, Scotland. II. Woodland vegetation and soils. *Journal of Ecology* **67**, 447–67.

Jeffrey, D. W. 1970. A note on the use of ignition loss as a means for the approximate estimation of soil bulk density. *Journal of Ecology* **58**, 297–9.

Kershaw, K. A. & J. H. H. Looney, 1985. *Quantitative and dynamic plant ecology*, 3rd edn. London: Edward Arnold.

Lambert, J. M. & W. T. Williams, 1962. Multivariate methods in plant ecology. IV. Nodal analysis. *Journal of Ecology* **50**, 775–802.

Lance, G. N. & W. T. Williams, 1968. Note on a new information-statistic classificatory program. *Computer Journal* **11**, 195.

Larcher, W. 1980. *Physiological plant ecology*. Berlin: Springer.

Macnaughton-Smith, P. 1965. Some statistical and other numerical techniques for classifying individuals. *Home Office Studies in the causes of delinquency and the treatment of offenders, No. 6.* HMSO.

Martin, M. H. 1968. Conditions affecting the distribution of *Mercurialis perennis* L. in certain Cambridgeshire woodlands. *Journal of Ecology* **56**, 777–93.

Martin, M. H. & C. D. Piggott, 1975. Soils. In *Hayley Wood: its history and ecology*, O. Rackham (ed.), 61–71. Cambridgeshire and Isle of Ely Naturalists Trust.

Melville, H. & B. G. Gowenlock, 1964. *Experimental methods in gas reactions*. London: Macmillan.

Noy-Meir, I. 1973. Data transformation in ecological ordination. I. Some advantages of non-centring. *Journal of Ecology* **61**, 329–41.

Noy-Meir, I., D. Walker & W. T. Williams, 1975. Data transformation in ecological ordination. II. On the meaning of data standardization. *Journal of Ecology* **63**, 779–800.

Orloci, L. 1966. Geometric models in ecology. I. The theory and application of some ordination methods. *Journal of Ecology* **54**, 193–215.

Orloci, L. 1978. *Multivariate analysis in vegetation research*, 2nd edn. The Hague: Junk.

327

REFERENCES

Pielou, E. C. 1984. *The interpretation of ecological data: a primer on classification and ordination*. New York: Wiley.

Pierik, R. L. M. 1975. Integrating photochemical light measurement, an ecological study in the Middachten woodland in the Netherlands. *Mededelingen van de Landbouwhogeschool te Wageningen* **65**, 1–16.

Piper, C. S. & J. A. Prescott, 1949. *Plant and animal nutrition in relation to soil and climatic factors*. British Commonwealth of Nations Scientific Organisation, 1949 Conference Proceedings.

Poore, M. E. D. 1955a. The use of phytosociological methods in ecological investigations. I. The Braun-Blanquet system. *Journal of Ecology* **43**, 226–44.

Poore, M. E. D. 1955b. The use of phytosociological methods in ecological investigations. II. Practical issues involved in an attempt to apply the Braun-Blanquet system. *Journal of Ecology* **43**, 245–69.

Poore, M. E. D. 1955c. The use of phytosociological methods in ecological investigations. III. Practical applications. *Journal of Ecology* **43**, 606–51.

Poore, M. E. D. 1956. The use of phytosociological methods in ecological investigations. IV. General discussion of phytosociological problems. *Journal of Ecology* **44**, 28–50.

Rackham, O. 1971. Historical studies and woodland conservation. In *The scientific management of animal and plant communities for conservation*, E. A. Duffey & A. S. Watt (eds.), 563–80. Oxford: Blackwell.

Rudeforth, C. C. 1970. *Soils of North Cardiganshire*. Memoirs of the Soil Survey of Great Britain, Harpenden.

Saunders, W. M. H. & A. J. Metson, 1971. Seasonal variation of phosphorus in soil and pasture. *New Zealand Journal of Agricultural Research* **14**, 307–28.

Stewart, V. I. & W. A. Adams, 1968. The quantitative description of soil moisture states in natural habitats with special reference to moist soils. In *The measurement of environmental factors in terrestrial ecology*, R. M. Wadsworth (ed.), 161–73. Oxford: Blackwell.

Streeter, D. T. 1974. Ecological aspects of oak woodland conservation. In *The British oak: its history and natural history*, M. G. Morris & F. H. Perring (eds), 341–54. Botanical Society of the British Isles.

Swan, J. M. A. 1970. An examination of some ordination problems by use of simulated vegetation data. *Ecology* **51**, 89–102.

Watson, E. V. 1968. *British mosses and liverworts*, 2nd edn. Cambridge: Cambridge University Press.

Whittaker, R. H. 1973. *Ordination and classification of communities*. Handbook of vegetation science, Vol. 5. The Hague: Junk.

Williams, W. T. & J. M. Lambert, 1959. Multivariate methods in plant ecology. I. Association-analysis in plant communities. *Journal of Ecology* **47**, 83–101.

Williams, W. T. & J. M. Lambert, 1960. Multivariate methods in plant ecology. II. The use of an electronic digital computer for association-analysis. *Journal of Ecology* **48**, 689–710.

328

REFERENCES

Williams, W. T. & J. M. Lambert, 1961. Multivariate methods in plant ecology. III. Inverse association-analysis. *Journal of Ecology* **49**, 717–29.

Williams, W. T., J. M. Lambert, & G. N. Lance, 1966. Multivariate methods in plant ecology. V. Similarity analyses and information-analysis. *Journal of Ecology* **54**, 427–45.

Wolfenden, E. A. 1979. *The synecology and autecology of selected woodland ground flora*. Ph.D. thesis, University of Wales.

Wolfenden, E. A., A. D. Q. Agnew, & D. R. Causton, 1982. An integrating photochemical light meter suitable for ecological survey. *Acta Oecologia, Oecologia Plantarum* **3**, 101–11.

Author Index

Italic numbers refer to text figures

Species Index

Italic numbers refer to text figures

332

Subject Index

Bold numbers refer to a main section which may extend over more than one page, and/or a definition. *Italic* numbers refer to text figures

abundance (value) 7, 58, 136, 146, 148, 155–7, 169, 191, 197, 201, 204, 213, 231, 239, 260, 305; *7.6*
acetic acid 48, 62, 64
aerial environment 10
photograph 13, 57
alliance 5
ammonium acetate 47–8, 61
ammonium phosphomolybdate 48
analysis of variance (ANOVA) 26, 27, 230, 239–44, 269; Tables 9.10, 9.18
angiosperm, 58, 66, 70; *see also* individual species
angle of rotation **159**
annual species 18
ant-hill 10, 51, 224
anthracene 63; Table 4.2
arch effect **207**, 209–13, 215, 218, 221–2; *7.26, 7.31*
aspect 10
association **5**, 52, 101, 104, 115, 145, 223, **239, 244**, 271, **279, 285, 305, 308, 310**; Tables 6.1–2, 9.1–5, 9.7–8, 9.10, 9.14, 9.16, 9.18–22; *6.1–3, 9.1–11*; *9.18, 9.23–36*
association between species **71, 72**, 82, 230; Table 5.1; **5.1**
Association Analysis 76, **98**, 105, 106–8, **112**, 115, **140**, 143–4; Tables 6.1–2, 6.11–12; *6.1, 6.3, 6.12*
average, *see* mean

background noise 42
bare ground 51
beta diversity **214, 215**; *7.30–5*
Betuletum 15; *2.2*
biennial species 18
binomial distribution 27
biomass 18; **20, 24, 27**
bivariate normal distribution 77, 89, 214, 230
Braun-Blanquet rating 8, 17; Tables 1.1–2

Bray & Curtis Ordination **173**; 181–2, 188, **215, 275**, 314; Tables 7.1, 7.5–6, 7.8; *7.15–18, 7.34, 9.16*
bryophyte, 66, 70, 287, 293, 314; *see also* individual species

calcareous boulder clay 52
calcifuge 70
Callunetum 50; *2.2*
Canonical Variate Analysis 243
case study **44**, 80, 115, 148, **244**; Tables 4.4–5, 8.1–3, 9.1–27; *4.1–7, 7.2–6, 7.11–12, 8.1–4, 9.1–38*
centring (of data) 157, **194**, 201, 202, 207, 215, 315, 318–19, 321–2; Tables 9.24–6; *7.7, 7.23, 7.31*
centroid 123–4, 126, 128, 157, 165, 239, 278, 282; *6.5–6, 6.8, 6.10*
Centroid Ordination *8.4, 9.17, 9.22*
chalk grassland 10, 130, 223; Plate 4
chance, *see* probability
characteristic root, *see* eigenvalue
characteristic species 42
characteristic vector, *see* eigenvector
chi-square (χ^2) 28, **72, 82**, 84, 88–9, 97, **98**, 107, 108, 112, 115, 140–42, 228–9, 316–22; Table 9.27
chi-square maximum (χ^2_{max}) **104**
class 5
climax adaptation number 169
clinometer 10
coenocline **213**, 214, 215, 222; *7.30–3, 7.35*
coenoplane **213**, 214, 218; *7.33–4*
combination (^nC_r) 98, 116, 140
compass 12, 13, 57; *7.16*
competitive species 41
component 159–66, 167–9, 193, 197–9, 201, 235; Tables 9.24; *7.23*
compression effect 208, 209; *7.25–6*
contingency table 28, **72**, 82, 202, 227–8, 229, 315; Tables 5.1–2, 9.27

337